RESIDENTIAL VENTILATION HANDBOOK

ABOUT THE AUTHOR

Paul H. Raymer is the chief investigator for Heyoka Solutions, which develops fully integrated housing solutions. He has been working with ventilation design and consulting for more than 30 years. Previously, he was president of Tamarack Technologies, Inc. and, prior to that, Weather Energy Systems, Inc. He has developed and brought to market more than 20 products, with patents and patents pending on several of them. He has taught ventilation, building science, air sealing, and combustion safety courses to diverse audiences, from home inspectors and weatherization specialists to Youth Build programs. He has taught the BPI Building Analyst course and courses developed by Conservation Services Group on air sealing and combustion safety, and has served as a technical consultant for many corporations including Dow Chemical, Rolscreen Corporation (Pella), Acorn Homes/Deck House, and APV Vent Axia. He is a member of the LEED for Homes IAQ TASC and has reviewed conference proposals for Affordable Comfort, the USGBC's GreenBuild Conference, and the ASHRAE Journal. He is a full member of ASHRAE (a voting member of the 62.2 SSPC) and the Home Ventilating Institute (chairman of the IAQ Committee), and is a HERS rater and a member of the Society of Building Science Educators. He also provides Manual J analysis for Energy Star Homes projects and is a member of the International Code Council.

RESIDENTIAL VENTILATION HANDBOOK

Ventilation to Improve Indoor Air Quality

Paul H. Raymer

New York Chicago San Francisco Lisbon London Madrid
Mexico City Milan New Delhi San Juan Seoul
Singapore Sydney Toronto

The McGraw·Hill Companies

Cataloging-in-Publication Data is on file with the Library of Congress

1 2 3 4 5 6 7 8 9 0 FGR/FGR 0 1 5 4 3 2 1 0 9

ISBN 978-0-07-162128-1
MHID 0-07-162128-8

Sponsoring Editor
Joy Bramble Oehlkers

Acquisitions Coordinator
Michael Mulcahy

Production Supervisor
Pamela A. Pelton

Editing Supervisor
David E. Fogarty

Project Manager
Jacquie Wallace, Lone Wolf Enterprises, Ltd.

Copy Editor
Jacquie Wallace, Lone Wolf Enterprises, Ltd.

Proofreader
Leona Woodson, Lone Wolf Enterprises, Ltd.

Art Director, Cover
Jeff Weeks

Composition
Lone Wolf Enterprises, Ltd.

McGraw-Hill books are available at special quantity discounts to use as premiums and sales promotions, or for use in corporate training programs. To contact a representative, please e-mail us at bulksales@ mcgraw-hill.com.

The pages within this book were printed on acid-free paper.

I gratefully dedicate this book to my loving wife, Kate, who has guided me through the years, children, and dogs, and put up with an addiction to fans, building science, and a continuing need to understand how things work.

CONTENTS

Chapter 14: Fan Types and Applications 14.1

Chapter 15: Special Application Ventilation 15.1

Chapter 16: Ventilation for Cooling 16.1

Chapter 17: Humidifiers, Dehumidifiers, Filters, and Ventilation Accessories 17.1

Chapter 18: Indoor Air/Environmental Quality Concerns 18.1

Chapter 19: The Future of the Residential Ventilation Arts 19.1

PREFACE

There are dozens of subjects one could study in this life, and it would seem that the old adage is true—"The more you know, the more there is to know." I remember clearly having absolutely no appreciation for how air moved through fans, ducts, or houses, thinking that it just happened. About 30 years ago I was asked to use the waste heat from the transmitters of a radio station to heat some offices, and that opened the door to the wonderful world of mechanical ventilation. I had to figure out how to move that air, what sort of fans to use, and how the air would move through the ducting.

To solve that problem I met a fan salesman named Jack "The Boze" Kennedy, who taught me about axial fans and venturis and the fact that fans can "suck" and "blow" and a bunch of other salesman stuff that is not the subject of this book. He also introduced me to a modular home company through which connection I met my wife, Kate.

Clearly a "house is a system." It is a working shelter for its occupants. That is its purpose—its reason for existence. People need to be warm, dry, and comfortable. And we need air to breathe. We can't just keep breathing the same air over and over any more than we can keep eating the same food over and over. This book is about ventilation for homes, both new homes and existing homes, and how to get the air to go where you want it to go and do what you need it to do. Remember that air doesn't always follow the arrows on the diagrams.

Paul Raymer

ACKNOWLEDGMENTS

As the building science doors have opened, revealing one layer after another of complexity and hopefully understanding, I have been honored to get to know and learn from many of the best minds in the business who have been instrumental in making this book a reality, who have contributed vast amounts of information to the science of air movement, and who I consider both associates and friends. My heartfelt thanks go out to the many reviewers of drafts of the book including Dave Wolbrink, Don Stevens, Steve Emmerich, Rick Karg, Jeff May, Terry Brennan, Judy Roberson, Rick Olmstead, John Harrell, Dennis Dietz, and Joe Nagan. I learn more from all of you every time I talk to you. I am exceptionally grateful to Vic Flynn of Panasonic who contributed a large number of the illustrations, Paddy Morrissey who drew them, and Dennis Livingston who guided me through the process. I will never remember all those who have educated me over the past 35 years, but I would like to recognize John Bower for his wonderful book on the subject of ventilation, and John Tooley who taught me how lazy air is, and Chris Clay, Bob Davis, Don Fugler, Paul Francisco, Tom Greiner, Dave Grimsrud, John Holton, Phil Kaluza, Mark Kelley, Steve Klossner, John Krigger, Joe Lstiburek, Mike Lubliner, Bruce Manclark, Neil Moyer, Collin Olson, Danny Parker, Duncan Prahl, David Price, Max Sherman, Armin Rudd, Doug Steege, Bruce Torrey, George Tsongas, Iain Walker, Eric Werling, and Jacki Golike and the folks at the Home Ventilating Institute, all of whom have taught me more than I could ever relay. I need to also recognize my non-building science friends who have put up with me testing and probing their minds and their homes—Steve and Jan Aubrey, John and Cate Dolen, and Ed and Kathy Furtek.

And last, but certainly not least, to my fantastic children—Abbey, Lizzie, and Adam and his wife, Heather, and grandchildren yet to be who bring into focus the reason why we struggle with all of this: the future.

Paul Raymer

Chapter 1
INTRODUCTION

The purpose of a residential ventilation system is to control odors and pollutants and indoor levels of moisture without causing discomfort to the occupants, unduly increasing the operating cost of the house, or adversely affecting the building envelope or the operation of other mechanical systems.

WHY A HOME NEEDS A VENTILATION SYSTEM

The instructor stood at the front of the room holding out a plastic bag and invited any member of the audience to come forward, place the bag over his or her head, and breathe comfortably for as long as possible. Although there were some chortles in the audience, not surprisingly there were no takers for the challenge. No one wants to stick his or her head in a plastic bag, but even if it had been a large (empty) bank vault at the front of the room, no one in the audience would have been willing to take up residence inside the vault with the door closed. We all know that we need air to breathe. It is obvious that a completely air sealed space is not good for human health. So why do we even have to pose a question like "Why does a home need a ventilation system?"

Homes are not air tight—but they're getting close to it. If a home were just a single story, air tight, empty box of air constructed of benign, non-volatile materials and there were no people in it, there wouldn't be an issue with ventilation. But houses are places to shelter people, and as soon as you move people into the box along with all their "stuff" you add air pollution. An adult human breathes approximately 66 pounds of air per day, approximately 880 cubic feet.[1] We breathe out carbon dioxide (along with a bunch of other stuff), and we breathe in whatever is in the air. If we don't change the air in the space, we end up breathing back in the stuff we breathed out.

1. Yberg, Ingvar, Indoor Air—The Silent Killer, Svensk Ventilation, p. 11, 2004.

A New York State study of a small random selection of homes built before 1940 averaged about 1.1 air changes per hour (ach).[2] In a larger study of new houses in New York State the average was only .23 ach.[3] The Northwest Infiltration Study[4] revealed that for a random group of houses built after 1982, some of which were designed to be "energy efficient," the average natural leakage rate was .4 ach, just above the current minimum ventilation standard of .35. To achieve the average, half of these homes had to be above the average and half below.

Most homes have more than people in them. They have all the things that people bring with them, like furniture and wall and floor coverings and cleaning products. These objects impact the indoor air evidenced by the new carpet or fresh paint smells. There are also the devices that we add for activities like cooking. The stoves, the food, and the cooking process all add to the interior chemical soup that we are brewing. Then there are the devices that keep us warm in the winter and cool in the summer and heat the water for the showers and the laundry. We try to vent the combustion appliances to the outdoors, but the air doesn't always flow in the right direction through the pipes. So sometimes those combustion products get added to the air in the living space. We park cars in the garage, and garages are commonly attached to houses and that air may get sucked into the house.

Then there is humidity—humidity from people, plants, cooking, combustion appliances, laundry, and showers. An average family of four generates two to three gallons of water per day from these activities.[5] In a house with one air change per hour, air moving through the house will carry away about 10 gallons of water. With that great an air change, the energy costs would be high and the house might feel too dry, dropping the winter relative humidity (RH) uncomfortably to less than 30 percent. If the air change rate drops to .25 ach, however, only two gallons of water will be carried out of the house. The house may feel damp, moisture will condense on cool surfaces (like windows), and mold is likely to flourish.

2. Rizzuto, J., <u>An Investigation of Infiltration and Indoor Air Quality</u>, NYSERDA 90-11, 1989.
3. Rizzuto, J., Nitchke, Traynor, Wadach, Clarkin, and Clark, <u>Indoor Air Quality, Infiltration and Ventilation in Residential Buildings</u>, NYSERDA 85-10,1985.
4. Palmiter and Brown, The Northwest Infiltration Survey, ECOTOPE, Aug. 5, 1989, Prepared for the Washington State Energy Office.
5. Bower, John, <u>Understanding Ventilation</u>, The Healthy House Institute, p. 55, 1995.

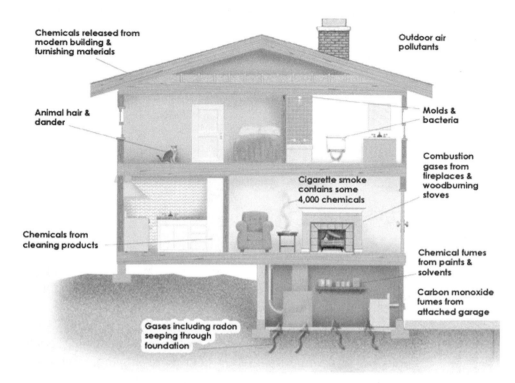

Chemicals released from
modern building &
furnishing materials

Outdoor air
pollutants

Animal hair &
dander

Molds &
bacteria

Combustion
gases from
fireplaces &
woodburning
stoves

Cigarette smoke
contains some
4,000 chemicals

Chemicals from
cleaning products

Chemical fumes
from paints &
solvents

Carbon monoxide
fumes from
attached garage

Gases including radon
seeping through
foundation

FIGURE 1.1   Indoor pollutants. (Panasonic/Morrissey)

It is clear that there are pollutants in the air in our houses and those pollutants are not good for us.

The only other reason <u>not</u> to use a mechanical ventilation system in a house would be if the air changes frequently enough naturally. Natural or passive ventilation strategies can work with the right designs in the right locations in the right conditions, but all of that is complex. For the air to move into and out of the house there need to be enough holes in the right places and there needs to be a reliable force to push the air into and out of those holes—infiltration and exfiltration. The general physics of this is covered in Chapter 7 of this book, but suffice it to say that these conditions don't happen reliably and the majority of people don't want a chilling breeze moving through their homes in winter.

So there are pollutants, we all need to breathe good air to stay healthy, and adequate air changes don't happen reliably or comfortably naturally. That means our homes need mechanical ventilation systems and that is what this book is about.

Mother's admonitions (the original building scientist):

- *It's a beautiful day. Go outside and get some fresh air!*
- *Open your window. Get some fresh air in that stinky room!*
- *Wipe your feet when you come in this house!*
- *Don't eat those cookies on the sofa! We'll get bugs!*
- *Go hang your pillows out on the clothesline![6]*

HOW TO USE THIS BOOK

People need to breathe. There have been thousands of conferences, studies, and education articles, papers, and books on ventilation in commercial buildings. Providing good air for people is an accepted and well documented concern.

Residential ventilation, however, is a much less certain subject. Although dozens of ventilation issues have been studied, conferences held, and papers written, it remains an area of remarkable contention. There are very few books on residential ventilation, and there are those who believe that purposefully moving fresh air through homes is not a concern. They believe that the volume of the house is too large and there is enough air moving in and out naturally through leaks and holes to satisfy any ventilation requirement.

Since a home is essentially a box of conditioned air that we contain with the walls, floors, and ceilings, we can keep the air in a range of conditions that is comfortable for the occupants. The better we build the box, the less energy it will take to maintain the right air temperature conditions. As we tighten the box, things we put into the box have an ever-greater effect on the quality of the air in the box. An architect who specialized in underground houses said that having eight people in his home was enough to heat it. He went on to say that if he got two of those people in a fight, he could send four people home. People generate about a hundred watts of heat and expel humidity and a lot of other things. There are chemicals and glues and gases that are emitted by the objects we bring into our homes that add together to create a chemical "mélange" in the air. Radon seeps out of the ground and out of our granite counter tops. Mold grows in places we can't always see. We park cars in our garages and store our paints in the utility closet with the air handler. There are

6. Mother.

combustion byproducts from gas stoves, fireplaces, furnaces, and water heaters. And we continue to tighten up our "boxes," our homes, to save energy, aiming to live more comfortably, physically and economically.

Admittedly the levels of most of these pollutants are generally small and appear to be tolerated by the majority of the population. The most obvious health effects are seen in the very young and the very old and people with heightened chemical sensitivities. But as we learn more about the effects of low-level exposure to many of these pollutants, we realize the value of a well designed, installed, and operated mechanical ventilation system in a home. We wouldn't want to keep reusing the same water in the shower over and over, and we definitely wouldn't want to keep re-breathing the air in our homes.

One fact is clear: as houses become tighter, reaching for better energy efficiency and comfort, the adage that "a house is a system" becomes ever more critical. Contractors and practitioners must understand the interaction between elements. If you burn the bacon, it's a natural reaction to turn on the range hood or open the windows to get rid of the smoke. Strong odors in the

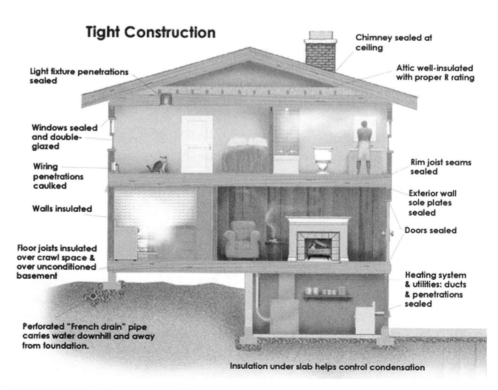

FIGURE 1.2   Tight construction details. (Panasonic/Morrissey)

bathroom beg for the activation of an effective bathroom fan. The object of both of these ventilating activities is to dilute the polluted air in the home with fresher outdoor air. Occupants generally only think about the exhausting side of the equation, accepting that the outdoor air will sneak back in from somewhere. But as all those cracks, holes, and leaks are sealed up to improve energy efficiency, the various exhaust systems in the house compete for the make-up air. It doesn't always enter by the best route, sometimes coming down through the water heater vent or down the fireplace chimney.

This guide is meant to be useful. It would be nice to think that an Oscar winning movie could be made from this book, but that seems less likely than a snake with elbows. It's just not the sort of popular subject that would cause many people to read it straight through from cover to cover. I have tried to provide a practical approach. If you are interested in installing or replacing a bathroom fan, for example, you should be able to get the information you need to select the proper fan, install it so that it operates correctly, and test that operation if you are so inclined and want to know if it is doing anything. If you are interested in just getting the job done, there are enough details to accomplish it. If you are interested in going into greater depth about indoor environmental quality (IEQ) or engineering, you will be able to find that information (or at least some of it) as well. (In the appendices there are a lot of great resources for going further.)

The early chapters of the book discuss fundamental design and sizing issues and describe system choices. The chapters flow on to installation issues and design and from there to installed system testing, and from there to code requirements, technical fan types, fan laws, and special ventilation applications like garages, radon ventilation, and ventilation for cooling.

One would think that since houses have been around for so long residential ventilation techniques should have been resolved long ago. The Romans were pretty good at it. The ancient Egyptians noticed that stone carvers working indoors had more respiratory problems than the ones working outdoors. They attributed it to dust.[7] But the fact is that we are still learning (or remembering what we have forgotten) about the subject. We are beginning to appreciate the relationships between all building components and systems and that leads to new information, new codes, and new standards. Although I have sought to use the most up-to-date information in putting this book together, things are changing even as I write it. The fundamentals are still fundamental (as the Romans and Egyptians understood), but how they are applied, the

7. Janssen, John E., "The History of Ventilation and Temperature Control," <u>ASHRAE Journal</u>, September, 1999.

rules for the their use, and the equipment for testing their performance are constantly changing. Luckily we now have the Internet where you can find the latest information. Hopefully this book will give you a roadmap of what to look for.

There is an expression that is commonly thrown about: "Build it tight. Ventilate it right." It's simple to say, but not simple to accomplish. Ventilating a house correctly is dependent on many variables, not the least of which is the climate. Products and technologies are changing. New answers and new questions are arising everyday.

There is a great deal of information about residential ventilation from conferences and technical papers, articles, and books. Much of that information and wisdom is referenced in this book. There is no single solution to the ventilation for a home that will fit every home in every climate condition with every occupant. If you bring a group of people into a room and ask them for their opinion on the condition of the air, you would certainly not get the same answer from every one of them. The American Society of Heating, Refrigerating, and Air Conditioning Engineers (ASHRAE) in its residential ventilation standard defines acceptable indoor air quality as, "air toward which a substantial majority of occupants express no dissatisfaction with respect to odor and sensory irritation and in which there are not likely to be contaminants at concentrations that are known to pose a health risk."[8] Such a condition is difficult to measure.

8. ASHRAE Standard 62.2-2007 "Ventilation and Acceptable Indoor Air Quality in Low-Rise Residential Buildings," p. 3.

Chapter 2
CHOOSING RESIDENTIAL VENTILATION SYSTEMS

Working with a large production builder, I noticed that the location of the toilet paper holder was carefully detailed on the plans, but there was no notation regarding the location or ducting design or control for the bathroom fan. "Everyone knows where the fan goes," he said. For a new house, designing the ventilation system *should* be as much a part of the design process as selecting the windows or the location of the rooms. It is much easier to visualize and design the whole ventilation plan as the house is being designed than it is to do it as an afterthought.

Making changes to the ventilation in an existing house is more difficult in the sense that the "system" wasn't planned in the first place. On the other hand, it is simpler to make modest improvements such as using a better bathroom fan, improving the ducting layout, or upgrading the control.

There are several different ventilation requirements in a home:

- Ventilation for people—this is the air people need to breathe to live healthy lives;

- Ventilation for combustion appliances—this is the air that is needed to have combustion appliances operate effectively;

- Ventilation for specialty purposes such as radon mitigation or garage exhaust;

- Ventilation for cooling—this includes elements such as ceiling/paddle fans, whole house fans, and attic exhaust fans.

All of these systems work together because everything in the house works together in one way or another. Other systems such as clothes dryers and central vacuums also draw on the air in the house and vent it to the outside. They should be included in adding up the total air moving through the house. (The window condensation in Figure 2.1 may indicate that the house is too tight.)

Some combustion appliances can be isolated from the home. There are combustion based space and water heaters (called sealed combustion) that draw in their own combustion air and vent their exhaust fumes to the outside. The air they use is completely isolated from the air in the house. Homes with

FIGURE 2.1 Window condensation. (Venmar)

sealed combustion appliances such as these don't have to include combustion air in their ventilation calculations as long as they are installed and working properly.

Some homes don't have combustion appliances—no gas range, clothes dryer, water heater, or fireplace. Obviously, if they aren't there, they don't have to be figured into the system calculation either.

If there is no attached garage, that would eliminate it from the ventilation system designs.

Perhaps surprisingly, homes with ducted heating and cooling systems need to consider those devices as part of the ventilation system. These systems are generally covered under the term HVAC or heating, ventilating, and air conditioning system. Most of the time they have been installed with little or no intentional connection to "ventilating," but it is very common for the ducting to be run through unconditioned spaces such as attics or crawl spaces and to have joints that are not effectively sealed. Those holes and gaps suck in or blow out air, pressurizing or depressurizing the home and inadvertently "ventilating" it. In the Pacific Northwest it was found that when the ducts running under the "bellies" of manufactured homes were carefully sealed, ventilation rates in the homes dropped to practically nothing, requiring the addition of mechanical ventilation systems.

Paddle or ceiling fans that are used to cool the occupants of the house by creating gentle, convective breezes commonly only move the air in a room and have no connection to moving the air into or out of the home. Whole house fans (or whole house comfort ventilators) that are used for cooling will purposely depressurize or suck the overheated air out of the house through open windows and vent it out through the attic to the outside. If these powerful fans are used at the wrong time or when the windows are closed, they will seek to draw the air from any other source including chimneys and will interfere with other components in the overall ventilation system.

The ventilation for people component can be divided into primary ventilation, which must run long enough to keep all the air in the home at an acceptable indoor air quality (IAQ) level, and "spot" ventilation that can be located at a "polluting" source such as a kitchen or bathroom that will draw out the worst of the contaminated air quickly, reducing the overall load on the building.

Smells from garbage that has been left to molder in a trash can undisturbed may not be immediately noticed. The smells are the symptom. Some or all of the items in the trash can are the cause as bacteria eats away at the contents. The longer the can sits there, the more the smells blend with the household air. This is a "spot" source. The cause of the problem can easily be removed from the house by emptying the trash can. The offensive odors in the air can be diluted with general household ventilation. Or if the can had its own exhaust fan, the immediate spot could be exhausted, removing the majority of the contaminant smells at the source before they had a chance to blend in with the rest of the air in the house.

CHOOSING VENTILATION SYSTEMS FOR A NEW HOME

The first step in properly deciding on the ventilation system for a new home is to consider all of the components that move the air in or out of the house— where they are located and how much air they move. Such a list must include the bath and kitchen fans, clothes dryers, central vacuums, and combustion appliances such as fireplaces, water heaters, and furnaces. (A gas clothes dryer is also a combustion appliance.) The system should also consider attic fans, whole house fans, radon fans, and garage exhaust fans. The fundamental object of this exercise is to provide a safe and comfortable condition in any room in the home in any weather condition. This summary must be assembled by thinking about the SYSTEM, including the house and all of its components.

Chapter 3 discusses how much air the ventilation for people system should move. Chapter 4 discusses the choices of the type of systems that are most applicable (see Figure 2.2). Bathroom and kitchen fans are considered "spot ventilation" fans since they are located close to spots or sources of pollutants—odors, smoke, and moisture.

When working with a new home design, all the options for ventilation are available. The ventilation component is as important or more important than the heating system or the type of windows or style of flooring. Good ventilation will keep the occupants healthy. It is difficult to quantify the impacts of low-grade pollutants such as carbon monoxide (CO) or formaldehyde (H_2CO) on people's health, but it is clear from the studies that have been done that keeping the air fresh in the house is an important component to people's physical and mental well being. The system can be as simple as dedicated exhaust fans with fresh air intakes into rooms or into the return side of the HVAC sys-

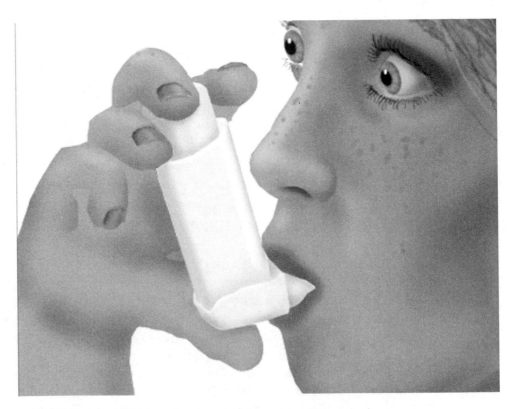

FIGURE 2.2 Bad IAQ and asthma problems are not a good mix. (Panasonic/Morrissey)

tem or as complex as a fully ducted supply or exhaust system, heat or energy recovery ventilator (HRV or ERV). New houses are generally much tighter and energy efficient than houses built in the past. We can no longer take for granted that breezes and drafts will be blowing through the rattling window frames or in through the basement or crawl space. Tighter houses are designed to trap the air inside. That air must be conditioned—heated, cooled, filtered, exchanged, humidified, or dehumidified—to keep the building and the occupants safe, comfortable, and healthy.

The primary ventilation system should be selected to gently and continuously change all the air in the house at a rate that will reasonably ensure the health of the occupants and the building. It should be designed in way that will provide air circulation and distribution through the all the rooms in the house, including the bedrooms. In still air, pollutants can accumulate around the occupants. If the air motion is excessive, the occupants will be bothered by drafts and discomfort. Tight houses are remarkably quiet, so the system has to be virtually silent. The quality of the motors in the primary ventilation system needs to be high because the system should be running continuously.

Since it is important to circulate the ventilation air around the people in the house, the rooms and hallways will likely serve as the ducting. Air may enter or leave through the mechanical system with grilles and vents located in strategic places, but the primary air passages will be the living spaces. There must be provisions for circulating air from room to room no matter how the building is being used, whether interior doors are open or closed. The house should be thought of as a large ducting system with the doors operating as dampers.

Outside air entering the system will be at a different temperature than the inside air. How will it be conditioned? Where will it enter so that it will not cause discomfort? One cubic foot of air moving into the house will be balanced by one cubic foot of air leaving the house.

A high quality bathroom fan can serve as the primary ventilation system if it is centrally located in a place where it can draw air from the house continuously most of the time. It shouldn't be in a master bathroom inside a master bedroom where the air will need to circumvent several closed doors.

Rule 1 of air motion: One cubic foot of air (cfm) moves into a house only if one cubic foot of air moves out of the house (1 cfm out = 1 cfm in). (This is easy to say and hard to believe.)

Particular thought should be given to how the air is going to get from the outside into the bedrooms and out of the bedrooms to the primary ventilation system. Houses last a long time and usually have numerous owners. Not all of them are going to open the windows at night in January when the temperature outside is 0°F!

A fully ducted, balanced ventilation approach such as a heat or energy recovery ventilator (HRV or ERV) will supply air to bedrooms and extract exhaust air from bathrooms via the ducting (see Figure 2.3). Air will still move effectively through the rooms from the inlets to the outlets. Conditioning will occur in the exchanger element. This is the ideal, no compromise approach.

Spot ventilation systems complement the primary ventilation system, taking the air out of the spaces that are the primary polluting sources—moisture in bathrooms and cooking byproducts in kitchens. By locating the fans close to the pollutant source, the fans can be less powerful and the pollutants are removed before they mix with the general air in the house and get circulated. These fans should be quiet enough so that they are used, but since the fans are likely to be operated intermittently, motor quality doesn't need to be as high as the primary ventilation system. In fact, in some bathrooms, a little fan noise is accepted as a good thing.

Any mechanical system will need to be serviced at some point and the ventilation system design should allow for access to the mechanical components so that filters can be changed and motors can be replaced. The systems should not be buried in walls or attics. The systems should be "commissioned" after installation to make sure that they work as they were designed. Airflows should be measured. It is surprising how many of these systems are not activated by the installers or designers to make sure that the ducting is all connected, as it should be, that filters are installed correctly, and that switches and controls are programmed.

The homeowner should not be expected to be a ventilation expert. They should be informed as to why the system was designed the way it was, how it is supposed to work, and how it should be maintained. Without that information, homeowners are likely to be concerned that the system is running too much and may try to defeat it and stuff the fresh air inlets full of socks to prevent air from leaking in or out.

After the primary ventilation system and the spot ventilation systems, other ventilation systems should be considered. It is much easier to install a radon mitigation system (and some jurisdictions require it) during construction. These systems will ensure that not only radon but also other soil gases

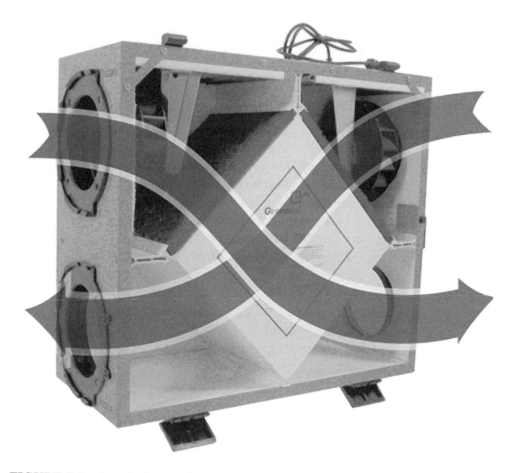

FIGURE 2.3   A typical cross-flow heat recovery ventilator (HRV). (Greentech)

are removed before they can enter the home. The pipe will run all the way from the slab at the base of the house through the roof. There should be a monitoring system that will alert the homeowner if the fan stops running.

Many green building programs and some codes require ventilating an attached garage. Ideally this should be done if there are no atmospherically vented combustion appliances in the garage that might be back-drafted by an exhaust fan. If the air pressure in the garage is lower than the air pressure in the house, any pollutants in the garage will "flow" toward the outside and not into the house.

> **Rule 2 of air motion:** Air always flows from higher pressure to lower pressure.

A whole house fan (whole house comfort ventilator) can be used for cooling the house when it is warmer in the house than it is outside. The traditional design is a large fan that sits in the middle of the top floor, drawing the air in through open windows and venting the house air into the attic and forcing it out through the attic vents. Some of these systems have motorized and insulated doors that block undesirable heat loss when the system is not in use. Others are ducted, removing air from multiple rooms and venting it directly to the outside, avoiding pressurizing the attic.

An attic exhaust fan vents the air from the attic directly to the outside. These fans will remove excess heat and humidity from the attic. The connection between the house and the attic should be carefully sealed for these systems to work effectively.

So for a new home the ventilation systems in order of importance that should be integrated into the design are:

1. Primary ventilation system
2. Spot ventilation systems (bathroom and kitchen fans)
3. Radon/soil gas ventilation system
4. Garage ventilation system
5. Whole house cooling fan (whole house comfort ventilator)
6. Attic fan

VENTILATION SYSTEM DESIGN FOR AN EXISTING HOME

The same system priorities apply to an existing house, particularly if it has been air sealed, reducing the natural infiltration rate. When a house reaches certain levels of tightness known as building tightness limits or BTL, mechanical ventilation is required to be certain that acceptable levels of air quality will be maintained.[1] (Note that this is really just a system check. More information on this process can be found in Chapter 7.)

---

1. Karg, Rick, "Survey of Tightness Limits for Residential Buildings," July 2001, for the Chicago Regional Diagnostics Working Group, www.karg.com/btlsurvey.htm.

Often in an existing home one is replacing an existing component like a broken bath fan or addressing a specific issue such as radon. If you know you want to deal with a radon problem, you don't need to spend a lot of time thinking about range hoods. You still need to keep in mind the system nature of the house. A radon mitigation system, for example, is likely to include a fan drawing the air out of the space under the slab. As long as the path from the ground to the space under the slab through the fan to the outside is a completely sealed system, there should not be any interference with other elements in the house. But if there is a sump pump in the floor of the basement or a crack in the floor that connects the basement air to the space underneath it, it is possible that air will be drawn from the house into the radon mitigation system. If air is drawn from the basement that is needed for the draft of the water heater, air could be drawn back down the chimney, and so on and so on. It is important to think through all the possibilities.

Replacing a component such as a broken or annoying bath fan should begin by evaluating the initial installation. The design of the existing ductwork should be checked for a straight and smooth path to the outdoors. The external hood should be checked for the function of its backdraft damper and whether any birds or other creatures have resided in the ducting. A bathroom fan exhaust duct that runs through an uninsulated space in a cold climate should be replaced with insulated ducting and possibly rerouted. It is common for significant moisture to collect in uninsulated ducting.

Before removing the offending device, try to determine its performance. This will give you an idea of the effectiveness of the existing installation and serve as a reference to recheck after the product has been replaced. Check to determine if any air is actually moving through the product. Hold a tissue up to the grille and see if there is enough airflow to pull it in. If it's a bath fan, try it with the bathroom door open and the bathroom door closed. (There are some testing methods described in Chapter 9.) Check the outside of the house at the grille with the fan running to see if the damper is being pushed open by the airflow.

Airflow is remarkably lazy. It is not eager to work. It will always seek the path of least resistance. It has to be pushed to get anything done.

Having made the initial observations regarding the installation of the existing product, refer to Chapter 6 to select, install, and test the replacement.

Rule 3 of air motion: Moving air will always seek the path of least resistance.

TOTALING UP THE VENTILATION IN THE HOUSE

The fundamental idea of ventilation rule number one is: One cubic foot of air coming into the house MUST be balanced by one cubic foot of air leaving the house. Don't be fooled into thinking that just because a fan is running and making noise that it is moving air. If you covered up one side of a "box" fan (one of those fans that people put in their windows in the summer), the fan blade wouldn't stop spinning, but the air would stop moving except a small amount in the neighborhood of the fan, stirred up by the spinning blades.

The normal reaction is to consider ventilation as being the exhaust fans, particularly the kitchen and bath fans, but as previously stated, there are other systems that impact the flow of air through a house. The infiltration and exfiltration of the air is driven by pressures inside and outside of the house, naturally moving the air. Then there are the chimneys designed to be exhaust systems to guide the exhaust air from combustion appliances (water heater, furnace, fireplace, etc.) out of the house. The clothes dryer is also an exhaust system, sometimes quite a powerful one. Central vacuums should take air out of the house as well. The mechanical systems for example:

Two small bath fans @ 50 cubic feet per minute (cfm) each	100 cfm
Range hood @ 400 cfm	400 cfm
Clothes dryer @ 200 cfm	200 cfm
Central vacuum @ 150 cfm	150 cfm
Total mechanical exhaust	850 cfm

This means that if all these systems were running simultaneously 850 cfm would have to be "leaking" into the house somewhere just for them to work. If the air doesn't come in at these rates, it won't be leaving at these rates no matter what the label on the fan box says. All of these mechanical devices will be competing for incoming air, and remember, air is lazy. It takes the path of least resistance. If the closest, biggest hole is a window, that's where the air is going to come from. If the closest, biggest hole is the fireplace chimney, that's where the air is going to come from (see Figure 2.4). If you close the bathroom door, the fan will seek to pull the air into the room from any opening—the crack under the door, the ceiling light fixture, the wall outlets, or the plumbing chases.

Besides the mechanical ventilation devices, there are the passive or natural ventilation devices like gas fired water heaters that rely on the "stack effect" or the natural effect of rising warm air to draw the combustion fumes up the flue. It doesn't take much for the mechanical systems to overcome the natural

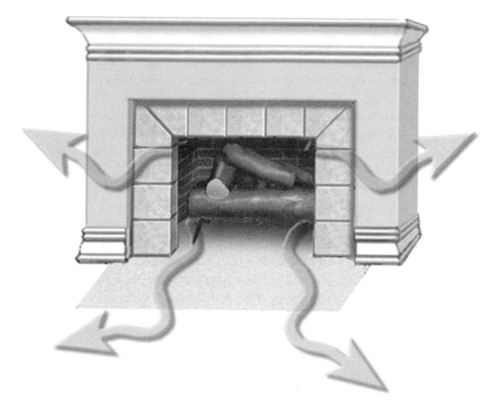

FIGURE 2.4   Fireplace downdrafting. (Panasonic/Morrissey)

systems and have the air flow the wrong way down the combustion exhaust pipe and into the house.

The goal is to know where the air is coming from and where it is going. It is much simpler and generally more effective to build a tight house with controlled and intentional airflows than it is to rely on natural air motion into and out of the house at the right times.

HOW TO READ A FAN BOX

There is some basic information printed on fan boxes that you find in the electrical products aisle that say more about the product than its manufacturer's name and the price. Understanding what the words mean is critical to selecting the right product.

Besides the brand and the bar code and the color of the grille is information like:

Airflow: 70 cfm: This means that the fan is capable of moving 70 cubic feet per minute (cfm) *under certain standard, prescribed conditions.* If the company has its products tested and certified under those prescribed conditions, the test laboratory's symbol should also be on the box like HVI (Home Ventilating Institute) or AMCA (Air Movement and Control Association). That airflow marking on the box will only provide an approximation of how the fan will perform once it has been installed. Remember: air is lazy. It doesn't like to push its way around corners or bump over the segments of flexible ducting or squeeze through small diameter ducting or push open the damper in the hood. All of those things resist the flow of air. Since every installation is different, the manufacturers had to agree on a common resistance or a test point so that all tested and certified products could be compared to each other. That test point was set low enough so that any product that is appropriate for a task (like ventilating a bathroom) could meet it. Since the test point is indicative of low resistance to the airflow, the installed performance will be lower (sometimes significantly lower) than what is marked on the box. Some manufacturers supply a second rating point, reflecting performance at a higher resistance. Some programs (such as Energy Star) require the fans to be tested and rated at a higher resistance. (Certified fan product manufacturers generate graphs of the fan performance at different resistances called "fan curves." These are sometimes included in the product literature and sometimes you have to ask for them.)

Not every fan box will include all this information, but here is some of the information that you might find:

Sound Rating: 4 Sones: A sone is a linear indicator of sound level. One sone is approximately the sound a quiet refrigerator makes in a quiet kitchen. Two sones are twice as loud as one sone. (More on sound in Chapter 5.)

For baths up to 65 sq ft, other rooms to 85 sq ft: This is another way of reflecting the airflow capability of the fan. The Home Ventilating Institute recommends eight air changes per hour for a bathroom to try to remove all the moisture that is generated from showers and other pollutants. If the bathroom has a floor area of 65 square feet and the ceiling is 8 feet high, the volume of the bath is 65 square feet times 8 feet or 520 cubic feet. Changing all that air eight times means multiplying the cubic feet by eight or 4160 cubic feet of air. Dividing that by 60 minutes comes out to 69.33 cubic feet per minute (pretty close to 70 cfm). Other rooms need fewer air changes (HVI recommends six air changes per hour in rooms other than bathrooms) because they have fewer polluting sources. Remember that the performance numbers on the box are based on the fan performing at a low resistance point. Also, remember that to achieve six or eight air changes in an hour the fan has to actually run for an hour.

Ceiling or wall installation: This is an indicator of how the backdraft damper in the fan is installed and product safety. If the damper is counter weighted for ceiling applications, it won't close properly if it is installed in a wall. (This may also indicate how the bearings in the fan motor are lubricated. Installing a ceiling-only fan in a wall may reduce the fan's life.) Fans that have been safety rated for installation in a wall have been tested to be sure that fingers cannot be inserted through the grille and impact moving parts.

UL: This indicates that the fan has been tested by a safety testing laboratory, in this case Underwriter Laboratories. (Other safety laboratories that do similar tests are ETL, CSA, and MET Lab.) These are privately operated companies that charge manufacturers to test their products according to established, standardized, safety test criteria.

HVI: This is the Home Ventilating Institute, a trade association for residential fan product manufacturers. They establish the test criteria for fans and related products so that the information on product performance can be easily and accurately compared. At this point, there is no obligation by the manufacturers to join HVI or test their products to these standards although some programs (such as Energy Star) require Certified performance results.

Armed with this information, it should be possible to select the right ventilation products, install them correctly, and test them to be sure they are operating properly. A good ventilation system will operate effectively for a long time, maintaining a livable level of indoor air quality and protecting the house that shelters the people.

Chapter 3:
BASIC APPLICATIONS, AIRFLOW, AND SIZING GUIDELINES

"Gentlemen know that fresh air should be kept in its proper place—out of doors—and that, God having given us indoors and out-of-doors, we should not attempt to do away with this distinction."—Rose Macaulay, British poet, novelist, and essayist (1881–1958).

The point of this whole process, of course, is moving the air around. If that's not happening, then not much ventilation is happening. It is important to have an understanding of how air moves in a house. Everyone seems to have a grasp on the concept that "warm air rises." Why is that? Is all air in the house the same just at different temperatures? If you don't use fans or open windows or turn on the HVAC, would the air just sit there? People talk about air changes. There is continuous debate among building science experts on what the rate of change should be. Does that mean all the air changes in the house? (That would be a rather spectacular process with all the air suddenly leaving the house to be replaced by a completely new batch!)

GENERAL AIRFLOW CONSIDERATIONS

Air is very interesting stuff. The air in our houses is made up of a variety of gases. "Standard air" consists of nitrogen, oxygen, argon, carbon dioxide, neon, helium, and other gases.[1] (It also contains a bunch of pollutants at varying levels.) Originally it all came from the outside. When the house was built, it was encapsulated, so to speak. When the windows are opened or the fans turned on, some of the old air is moved out and new air is drawn in somewhere else. The volume of airflow is measured in the U.S. in cubic feet per minute or cfm. (In other parts of the world it is measured in cubic meters per second or cubic liters per hour.) Imagine an invisible box—a box of air, 1 foot by 1 foot by 1 foot, floating there in front of you. Because it's in the same room, it's about the same temperature and full of about the same particles and

1. http://www.engineeringtoolbox.com/air-composition-d_212.html.

The advent of the "blower door" has given us a tool for determining a close approximation of what the natural air change rate of a home is. The blower door is a large fan that is temporarily installed in an outside door of the house exhausting air (see Figure 3.1). The flow rate is increased to a point that overcomes any natural air transfer, and the resulting number can be adjusted to reflect a natural air change rate indicating where the house can be tightened up and how much mechanical ventilation might be needed.

gases that you are breathing. There are a finite number of air molecules in the box. If the temperature should incresae, the air molecules would expand and fewer of them would be able to fit into the box, the box would get lighter, and it would move upward. If the temperature in the box went down, more air molecules could fit in the box, the box would get heavier, and it would sink. Or think of a hot air balloon filled with warm air light enough to lift the balloon, basket, and occupants up into the sky and float on the winds.

An air mass carries water-vapor molecules. If the number of molecules of air in the invisible, 1 cubic foot box goes up (because the temperature goes down), the amount of moisture relative to the amount of air would also go up. The amount of water-vapor the cubic foot of air can hold is dependent on the temperature—more water-vapor can exist in a space if it is warm than if it is cold. In the summer, there may not be many air molecules in the cubic foot box, but with a high relative humidity (RH), it can be saturated because the amount of moisture relative to the number of air molecules may be high. In the summer when it is warm, the cubic foot of air may hold a lot of water-vapor molecules and be saturated because the amount of moisture relative to the amount of air is high. One hundred percent relative humidity means that the air is totally saturated. It can't hold any more water, creating the probability of rain. At high levels of RH, the air can't carry away the moisture generated by our bodies and we feel more uncomfortable. At a point where warm, moist air touches a cooler surface, a surface that is cooler than the dew point temperature, the moisture in the air will condense on that surface, changing from a gas to a liquid. The dew point can occur on the surface of grass in the morning or somewhere inside a wall.

There are a lot of other things in a cubic foot of "air" that can upset the make-up of "standard air"—other gases like carbon dioxide (which we breathe out), carbon monoxide (from burning things or cars in the garage), ozone, formaldehyde, and other chemicals of daily life. And there are lots of particles, both small and large, floating around in the air.

Residential ventilation strategies are based on the concept that the outside air is generally much cleaner than the inside air. Mothers say, "Go outside and get some fresh air." There are times when that isn't true, when we are actually advised to stay indoors, but residential ventilation strategies make the assumption that the polluted air is in the house and needs to be changed with outside air and most of the time in most places this is true.

A thermostat only "knows" that when it gets cold, the heating system can be turned on and heat will be produced. We assume the outdoor air is better than the indoor air.

When the air in the house is changed by the ventilation system, not all the air gets changed at once. Each "complete" air change will reduce the pollutant level by one half. Imagine a large container partially filled with a cup of black coffee. If a cup of clean water is added, the liquid in the container is now half as "polluted." Empty one cup of liquid out of the container and add another cup of clear water, the liquid is now one quarter as polluted as it was at the beginning, and so on and so on. As long as no new "pollutant" or cof-

FIGURE 3.1   A blower door (PHR).

The U.S. EPA has a Website that provides an indication of the outdoor air on a daily basis at www.AirNow.gov. For Canada there is the Air Quality Health Index www.ec.gc.ca/cas-aqhi that provides daily outdoor environmental conditions for various communities across Canada. There are times and locations when the outdoor air should be treated with caution, particularly for people who are susceptible to compromised conditions.

fee is added to the container, the liquid will become increasingly less polluted. Because the amount of fresh liquid is equal to the amount of "polluted" liquid, this is equivalent to a 50 percent change rate. "The solution to pollution is dilution."

In terms of air, if the air changes in the house are continuous, the air will stay generally acceptable. If the natural air changes (what is called air changes per hour or ACH) are too great, the air may be fresher but the energy cost will be high. If the ACH is too low, the air in the house will be stale and possibly unhealthy. The trick is to find the balance between the two. There are Standards and suggested rates (see Chapters 12 and 13), but a great deal depends on the house, the location, and the occupants. Some people like to open their windows throughout the year. Some people never open their windows. Some people live in climates where natural breezes are welcomed. Some homes are in places where everything possible must be done to keep the outside . . . outside.

FAN SYSTEM CHOICES

The most efficient location for exhausting the air from a home is to remove it from the places where the most pollutants are generated, like the bathrooms, kitchen, garage, and from the ground under the slab. Fresh air should be introduced into places where people spend the most time like the bedrooms, living room, family room, or TV room. The most common fans are fans that mount in the ceiling like a typical bath fan and fans that are mounted in the range hood. Paddle fans hanging from the ceiling have come back into fashion in recent years, helping to stir up the air in both summer for cooling and in winter for pushing the warm air back down (see Figure 3.2). Window or box fans, pedestal fans, oscillating fans, and hassock fans are used for cooling people by generating a breeze or pushing warm air out of a hot house. Specialty down-drafting range fans have gained popularity.

FIGURE 3.2   A ceiling fan (Broan).

There are other, less familiar fan choices. There are "in-line" fans that are used to pull and push air through a duct. They can be mounted remotely from the space they are venting with the advantage of removing the fan noise from the space. As "work-horse" fans, they can be used for hauling radon or soil gas-polluted air from under a slab or increasing the length of the duct run from a dryer.

There are large, propeller fans that are used as cooling systems in homes (see Chapter 16). These are called "whole house comfort ventilators," although they are more commonly referred to as "whole house fans." Propeller fans are also used as attic fans, mounted in the roof or gable end of the attic and pushing the hot air out.

Of course, there is the blower in the air handler that pushes the heated or cooled air around through the ducts to condition the space. These are generally the most powerful and energy consuming fans in the house as they are designed to move the most air, but they do circulate air throughout the house and, coupled with a fresh air intake, can be used as the primary ventilation system.

Different ventilation components have different jobs to do and should be sized and selected in response to their purpose. Although some ventilating tasks can be combined, in most homes multiple systems will be required. Always keep in mind, however, that "the house and the occupants are a system."

AIRFLOW CONSIDERATIONS FOR THE PRIMARY VENTILATION SYSTEM

The bathrooms and the kitchen are known sources of moisture and other pollutants. Those pollutants can be exhausted immediately from the "source" by exhausting those points when the bathrooms or kitchen are being used, which is why this is called "spot" ventilation. But the people and the materials in the house produce other, more widely dispersed pollutants that also need to be exhausted. Primary or whole house, air quality ventilation systems are dedicated to improving the air quality of the whole house. These systems can be run continuously or intermittently as long as they are run frequently, at least for a part of every hour.

Permanent materials in the house such as cabinets and flooring continue to "out-gas" long after they have been installed. Carpeting and furniture have chemicals that are added to the air. Cleaning materials, paints, deodorants, pets, and people all continuously add to the gases and particles in the air. Some of these sources occur intermittently. People might not be in the house all the time. Some of the pollutants increase and decrease throughout the day. The primary ventilation system needs to deal with all of that in order to keep the air acceptable for the occupants of the house.

ASHRAE 62.2-2007 defines acceptable indoor air quality as "air toward which a substantial majority of occupants express no dissatisfaction with respect to odor and sensory irritation and in which there are not likely to be contaminants at concentrations that are known to pose a health risk."[2] These are subjective measurements, but there are certainly objective measurements in indoor air that need to be controlled as well. By running a primary, whole house air quality ventilation system regularly or continuously the level of the pollutants in the house is not allowed to build up and become uncontrollable.

Primary Ventilation System Sizing

A great deal of work has been done to try to determine what that optimum, continuous ventilation rate should be. Calculating the requisite airflow of this system is not as simple as figuring out what size fan to put in the bathroom or kitchen (see Figure 3.3). This is one of those occasions where "one size does not fit all." One airflow number or fan size is not going to satisfy all the local conditions—house size, occupancy, climate, etc. There are a number of

2 ASHRAE Standard 62.2-2007.

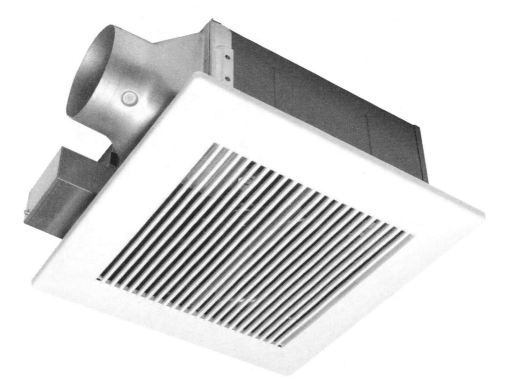

FIGURE 3.3   A ceiling exhaust fan (Panasonic).

approaches that can arrive at a number "in the ball park." The flow rates vary from 30 cfm for a small, one bedroom house to 165 cfm for a large, seven-bedroom house. If the flow rate is too small, its ventilating effects will be inconsequential. If the flow rate is too large, it will put an unnecessary cost burden on conditioning the air. Occupants have different sensitivities to air quality conditions. They bring different materials into their homes. They participate in different activities for different lengths of time. And they live in different places, with different weather conditions in homes that have been constructed differently, with different levels of tightness. On top of all that, the installed performance of systems will vary.

Ideally the ventilation rate would be determined by detrimental health effects that have been *avoided* or the number of diseases that *didn't* happen because of adequate ventilation. Although there have been a number of studies on the relationship between health, indoor air quality, and ventilation, it is

Note that these flow rate numbers are rates of airflow only. They do not refer to DIRECTION or TYPE of airflow. The direction of flow will affect whether the house is likely to be positively or negatively pressurized and that will have a major effect on moisture and building durability. See Chapter 7 on House Pressures.

As a general rule of thumb, if the home is located in a hot and humid climate and relies on air conditioning to keep the living space comfortable, a positive ventilation approach is preferable, pushing outside air into the house.

only the most extreme cases that can be quantified. The effects of the myriad of everyday, low-level pollutants on the health of the home's occupants are virtually unknown. We spend about 80 percent of our lives indoors and know that levels of chemicals such as formaldehyde, chloroform, and styrene are two to 50 times higher indoors than outdoors.[3]

Odors are the one thing that will prompt people to operate ventilation systems and windows. If the ventilation rate drops below 5 cfm per person for more than a couple of hours, occupants will be bothered by odors.

On the other hand, as ventilation rates increase so does the cost of conditioning the air. The impact of energy costs fluctuates with the interest in tightening up homes and the increasing cost and declining availability of heating fuels.

One approach to flow rate sizing is to use the number of bedrooms as the guiding factor, estimating that the number of bedrooms is a reasonable guide to the number of people that will be living in the house, estimating two people in the master bedroom and one in each additional bedroom. ASHRAE proposes that each person should be allowed 15 cfm. So a two-bedroom house would require roughly 45 cubic feet of air moving through it continuously.

A second approach is to design the ventilation system to provide 0.35 air changes per hour. A 1500 square foot house with 8-foot ceilings would require a 70 cfm fan. (Another approach to this is to use 5 cfm per 100 square feet of floor area, which in the case of this 1500 square foot home would suggest 75 cfm. Note that this approach may provide a rate that is excessive in large houses.)

3. http://articles.latimes.com/2006/jun/22/local/me-teenair22.

A third approach is to multiply the floor area by 0.01 and add to that 7.5 times the number of bedrooms plus one (for the extra person in the master bedroom). A two-bedroom, 1500 square foot house by this calculation would require a 37 cfm fan. This calculation assumes that the house will be leaky enough to allow 2 cfm/100 square foot of floor area to be naturally added to the fan flow rate, in this case 1500/100 × 2 = 30. Adding that to 37 cfm of mechanical ventilation provides 67 cfm, very close to the other calculations. If the house is built tightly, does not have a conditioned air HVAC system, or it is a time of the year when that system isn't running, it is not likely to be that leaky. To be sure of adequate ventilation, it is preferable to size, design, and install the mechanical ventilation system to handle the whole load.

ASHRAE offers both a formula based on the floor area of the house and the number of bedrooms[4] as well as tables that can be used without doing any math (see Tables 3.1a and 3.1b).

TABLE 3.1a   Ventilation Air Requirements, cfm *(from ASHRAE 62.2-2007 Table 4.1a [I-P]).*

Floor Area (ft2)	Number of Bedrooms				
	0-1	2-3	4-5	6-7	>7
<1500	30	45	60	75	90
1501-3000	45	60	75	90	105
3001-4500	60	75	90	105	120
4501-6000	75	90	105	120	135
6001-7500	90	105	120	135	150
>7500	105	120	135	150	165

TABLE 3.1b   Ventilation Air Requirements, L/s *(from ASHRAE 62.2-2007 Table 4.1b [SI]).*

Floor Area (ft2)	Number of Bedrooms				
	0-1	2-3	4-5	6-7	>7
<139	14	21	28	35	42
139.1-279	21	28	35	42	50
279.1-418	28	35	42	50	57
418.1-557	35	42	50	57	64
557.1-697	42	50	57	64	71
>697	50	57	64	71	78

4. ASHRAE Standard 62.2-2007

For square foot measurements:

$$Q_{fan} = 0.01A_{floor} + 7.5(N_{br} + 1)$$

where

Q_{fan} = fan flow rate in cfm (cubic feet per minute)

A_{floor} = floor area in square feet

N_{br} = number of bedrooms (not less than one)

For metric measurements:

$$Q_{fan} = 0.05A_{floor} + 3.5(N_{br} + 1)$$

where

Q_{fan} = fan flow rate in L/s (liters per second)

A_{floor} = floor area in square meters

N_{br} = number of bedrooms (not less than one)

Using the table is pretty straightforward: select the floor area and the number of bedrooms and the intersection is the amount of airflow required for the primary ventilation system. A 2000 square foot house with two bedrooms, for example, would require 60 cfm of airflow. (ASHRAE 62.2 will be discussed in more detail in Chapter 13.)

BATH FANS

Besides the furnace blower, the most common fan in the house is the bath fan. It is surprising to find bathrooms that don't have fans. There has been a long battle about whether or not every bathroom should have a fan mandated by code. Some of that conflict was generated by a definition of the purpose of the bath fan. For many people the purpose is to provide "masking noise," not to move the air. There are still those who think that if there is a window in the bathroom, a fan isn't necessary. There are many reasons not to open a window in a bathroom—from climate to neighbors. And even if the window is open, it doesn't mean the air will move effectively through it (particularly if the bathroom door is closed). So for many years bathroom fan manufacturers weren't particularly concerned about how noisy the fans were. In fact, it was almost considered, "the noisier the better."

Fans are sometimes run on their own wall switch, but more commonly the same switch that controls the lights operates them (see Figure 3.4). To make

FIGURE 3.4   A bath fan with a light (Panasonic).

that even easier, fan manufacturers added the lights to the fan so the customer bought everything in a single package. By having the fan come on when the light turned on, it was thought that there would be at least some mechanical ventilation in the bathroom (and it made installation simple). The occupant could avoid using the noisy fan only by functioning in the dark. The light switch was a crude "occupancy sensor."

The fan is mounted in the ceiling so the light is in the ceiling and to facilitate the duct running from the fan to the outside.

These ceiling mounted fans are usually a complete package with the fan, the light, the grille, the mounting system, and the duct connection with a backdraft damper to keep the cold drafts from coming back into the room from the outside. These dampers rely on air motion to open and gravity to close, which is why fans designed for ceiling mounting should be installed in ceilings. If the bathroom design calls for the fan to be wall mounted, the fan description should state that the fan is designed for wall mounting.

As houses became tighter during the energy disruptions in the early 1970s, building scientists recognized that ventilation played an important role in occupant health and comfort and building durability, and it seemed obvious that the bath fan could be used for more than masking the noises in the bathroom. The fan could be used to improve air quality in the bathroom and the house, but it would need to be run longer and to do that it would need to be quieter.

The bath fan doesn't have to actually be in the bathroom. The goal is to get the bad air out of the room and to the outside. So the fan can be remotely mounted with just a grille in the bathroom. "In-line" fans are ideal for this sort of installation. They can be installed virtually anywhere in the duct line, removing their operating noises from the bathroom and allowing for an array of grille choices in the bathroom. The grilles can be mounted in the ceiling or wall, high up or low down. They can be mounted in a shower area or behind a toilet. They can be mounted in the frame of the mirror to keep the fogging to a minimum. The important thing is that the opening in the bathroom should match the capability of the fan to draw air through it.

Fans can be mounted at the end of the ducting on the outside surface of the home. It is important that in such installations a backdraft damper be added to the ducting, preferably near the fan, to prevent unconditioned outside air from flowing back into the house when the fan isn't running. The fan itself may not include the damper although one may be located in the external hood.

There are also fans that can be mounted in an exterior wall of the bathroom, venting directly to the outside (see Figure 3.5). Some can be mounted in the glass of a window. These are less common with the difficulty of cutting a round hole in a piece of insulated glass. The advantage of these direct exhaust applications is that there is no ducting to be concerned with and the installed performance should equal the rated performance. A disadvantage is that the through-wall opening is bigger than the opening for a ceiling mounted fan and the closure system has to be more positive to prevent air from flowing back into the house.

Bath Fan Sizing

The close proximity of the bath fan to the source of moisture means that it can be smaller than it would have to be otherwise. There are a number of considerations in determining the "size" or rate of airflow for a bath fan. The Home Ventilating Institute (HVI) recommends a bath fan be capable of changing all the air in the bathroom eight times an hour. So if the bathroom is 8 feet by 5

FIGURE 3.5   A through-wall exhaust fan (Panasonic).

feet with an 8-foot ceiling: $8 \times 5 \times 8 = 320$ cubic feet, to change that air eight times in an hour multiplies the volume by eight, or 2560 cubic feet of air, to move each hour. Since there are 60 minutes in an hour, dividing 2560 by 60 results in a 42 cfm rated fan. (Another way to calculate this is to allow for 1 cfm per square foot. So this 40 square foot bathroom would need a 40 cfm fan. Just about the same.)

For a larger bathroom (over 100 square feet), HVI recommends sizing the ventilation by the appliances.[5] (See Table 3.2).

For example, if the bathroom had a toilet, a separate shower, and a bathtub, it would require $50 + 50 + 50$ or 150 cfm. These areas could be vented separately with separate fans or with a single, remote fan with multiple inlets or with a single, 150 cfm ceiling fan.

5. http://www.hvi.org/bguide.html.

TABLE 3.2 Recommended airflow per bath appliance.

Toilet	50 cfm
Shower	50 cfm
Bath tub	50 cfm
Jetted tub	100 cfm

ASHRAE in Standard 62.2-2007 offers two numbers for bathroom fans—20 cfm if the fan is to be operated continuously and 50 cfm if it is to be operated intermittently.

It is important that the fan run long enough to accomplish a reasonable air change. That is one reason why HVI suggests higher airflows. If the 42 cfm fan is on for only 20 minutes in the 320 cubic foot bathroom, it will only accomplish about two and a half air changes. A more powerful fan will be able to change the air in the bathroom more quickly and will not need to run as long as a fan with a lower cfm capability. Of course, exhausting more air from the house has an energy penalty to it, but that is generally small. (For a fan that moves 150 cfm running for 20 minutes when it's 30°F outside and 70°F inside, the energy required to heat the air is about 2160 Btus or $0.004 (at $0.21/kwh). Remember though that the bath fan is just one part of the total ventilation inventory in the house, and as its exhaust airflow increases the amount of make-up air needed also increases (remember rule 1 in Chapter 2: 1 cfm out = 1 cfm in).

It is also important that the installation of the fan be in compliance with the manufacturer's specifications. The manuals that come with these fans are pretty good at offering installation details, but who reads installation manuals? Long, twisting runs of flexible ducting will greatly reduce the actual installed performance of the fan. Air is lazy.

KITCHEN EXHAUST FANS

Range hoods are designed to trap or "entrain" the fumes and smoke and moisture from cooking and sweep them out of the room before they pollute the rest of the house. Kitchen exhaust fans are commonly located or associated with a range hood that is mounted over the stove. Some hoods have the fan installed in the hood and some have the fan installed remotely (see Figure 3.6). In an attempt to ease installation, some hoods are designed to simply

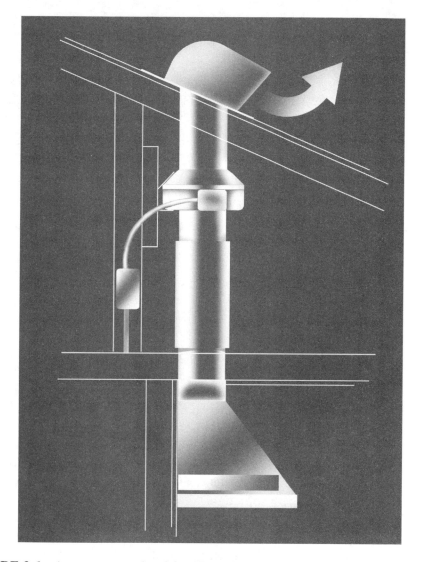

FIGURE 3.6 A remote range hood fan (Fantech).

draw the air in, pass it through a "filter," and then vent it back out into the kitchen. The filter in the hood is designed to trap the grease and an odor "absorbing" material such as charcoal or carbon is designed to trap the odors. Grease will coat the carbon, making it ineffective so these "traps" need to be replaced regularly. Because these fans do not vent to the outside their effectiveness is limited.

Backdrafting is an important consideration in sizing any fan but particularly a kitchen range hood because of the potential power. A casual over-sizing of the fan in order to satisfy the "bigger is better" need can lead to potentially disastrous consequences. Imagine a winter party with a cozy fire in the fireplace. The host flips on the range hood fan when he burns some hors d'oeuvres and the house fills with smoke from the back drafting fireplace. At the same time the gas water heater in the basement backdrafts, drawing in carbon monoxide and other combustion by-products. As houses become tighter it takes less and less negative pressure to cause problems with combustion appliances seeking to use the same air. It is best to use a less powerful fan and test the performance to make sure.

As an alternative to the range hood mounted fans there are "down-drafting" vents that pull the air down and out. There are also fans that are used as "general" kitchen fans, exhausting through a wall or window or from the ceiling of the kitchen.

There has been a trend toward installing commercial appliances in the kitchen under the "bigger and more powerful is better" philosophy. With commercial stoves installed, the inclination is to employ commercial range hoods as well, making a significant statement about size and prestige and power. In commercial applications there are strict code requirements regarding airflow and make-up air. Commercial hoods need an adequate amount of airflow power to draw the air through the filter system at a high enough velocity to entrain the grease and other particulates and limit their settling on other parts of the kitchen and serving as a source for grease fires. A higher airflow velocity will keep the particles suspended and allow them to be vented out of the kitchen. The Uniform Mechanical Code (UMC) requires make-up air for commercial kitchens, but with few exceptions there are no such requirements in residential codes. However, if a commercial hood is used in a residential application, the commercial codes should be applied in order for the product to work properly.

TABLE 3.3 Recommended sizes for range hoods.

HVI	40 to 100 cfm per linear foot of cooktop	For hoods located against a wall
	50 to 150 cfm per linear foot of cooktop	For "island" hoods
ASHRAE 62.2-2007	100 cfm (50 L/s) intermittent	If operated continuously, it can
	Five air changes per hour continuous	be all or part of the primary ventilation system

Kitchen Fan Sizing

HVI recommends between 40 and 100 cfm per linear foot of cooktop if the hood is mounted next to a wall (see Table 3.3). If the hood is located over an "island," the recommended rate is increased to 50 to 150 cfm per linear foot (see Figure 3.7). By this criteria a typical four burner, 30-inch wide range would require a 100 to 250 cfm exhaust fan. Commercial criteria commonly call for 300 cfm per linear foot. This would indicate the need for a 750 cfm exhaust fan. Although there are limited code requirements regarding make-up air, remember that 1 cfm out equals 1 cfm in. For the larger, commercial sized range hood fans to work properly they will require some means of supplying make-up air whether or not make-up is a local code requirement (see Chapter 8).

ASHRAE 62.2-2007 requires 100 cfm of venting from the range hood if there isn't a kitchen exhaust fan running continuously that is capable of five air changes per hour. For example, a 300 square foot kitchen would require a 200 cfm fan exhausting continuously.

There is another sizing suggestion of 1 cfm per 100 Btu rating of the cooktop. This would put the commercial cooktops in a completely different sizing

FIGURE 3.7   A range hood (Air King).

category since they are rated in excess of 10,000 Btus, some as high as 280,000 Btus. The thing to remember is that hopefully all those burners won't be operating at the same time. Using the linear approach for sizing the fan (meaning a fan of no more than 600 cfm) will be adequate in the vast majority of applications.

The hood should be at least as wide as the cooktop—preferably an inch or two wider on each side and located between 2 and 3 feet from the cooking surface. It should be simple to maintain. Some hoods have fan blades that can be removed and washed in the dishwasher. The grease filters should also be cleaned regularly.

OTHER VENTILATION SYSTEM SIZING

Other ventilation systems will be discussed in more detail in Chapter 15, but they should be considered as a part of the total ventilation "inventory" in determining the total airflow. Radon or soil gas mitigation systems suck the air from under the basement slab or crawl space floor and from around the foundation walls and blow it to the outside. These systems should run continuously but they draw very little air from the inside of the house except through cracks or gaps in the floor.

Attached garages can also add pollutants to the house. If the pressure in the house is lower than the pressure in the garage, air will flow from the garage to the house carrying with it all the pollutants that are in the garage air. A garage exhaust fan can mitigate this problem, but care must be taken to be sure that a fan blowing out of and depressurizing the garage will not effect any combustion appliance located in the garage. The right airflow for a garage exhaust fan will vary by the tightness of the garage and by the run time. If the garage is very tightly constructed and the fan is run continuously, a very small airflow will depressurize it—flows as small as 20 cfm. A typical garage with intermittent operation may need as much as 100 cfm per bay.

Chapter 4
SYSTEM DESIGN CHOICES

The best point to select a ventilation system design is when the house is being designed, when the ducting, wiring, and plumbing are being laid out. At that time all the options are open and there is an opportunity to carefully think through how the systems (and their installers) will work together. It is the time to think about how the systems will first be used and how they might be used after the first occupants leave and another family moves in. It is a time to think about what might happen in the future, both in terms of modifications to the house, and also in terms of maintenance of the systems. Houses have been known to last for hundreds of years. It is the best use of the materials. Parts of the house that will last for 200 years should not have to be ripped out to repair or replace parts that will only last 10 or 20 years. If the ventilation system is not designed properly in the first place, it is likely to shorten the life of the building. If poor equipment is used, it may need to be serviced regularly or replaced sooner. If good equipment is purchased and a good system is designed but it is installed poorly, the system won't be effective, and it will need to be serviced. The unfortunate fact is that any mechanical system no matter how beautifully designed and manufactured will need to be repaired or replaced at some point. Burying fan systems in wall or ceiling cavities with no access guarantees that someone in the future is going to have to rip that wall or ceiling apart for even a simple adjustment or repair.

This certainly doesn't mean that an existing ventilation system cannot be repaired or replaced or improved in an existing house. It's just more of a challenge. Running ducting through existing walls and ceilings is not as simple as snaking a wire (and even that can be a challenge at times).

The options are essentially based on where and how the mechanical components of the system are placed. Pollutant source control requires limiting the sources as a first step, keeping the materials that contain significant, harmful chemical components out of the building—like formaldehyde and carbon monoxide. Radon and soil gas mitigation fans and bath and kitchen fans are located near the source of pollutants and provide the activity related ventilation control. They do not have to be very powerful because they are located near the polluting source, the distance the air needs to move is relatively small, and, like a local whirlpool, they will suck in what is around them. These devices are part of the ventilation system, but, as discussed in the previous chapter, they need to be accompanied by a whole house, air quality, primary ventilation system.

EXHAUST-ONLY PRIMARY VENTILATION SYSTEMS

An exhaust-only system locates the mechanical component on the exhaust side, like a bathroom fan, blowing air out of the house. The exhaust-only approach depressurizes the house, sucking the air out of it, relying on leaks and cracks and gaps in the building shell to let the air back in (see Figure 4.1). The exhaust fan may be doing double duty and located where pollutants are generated like in a bathroom, or it may be centrally located in a hallway. The inlet air "leaks," however, will be spread about the house and may not effec-

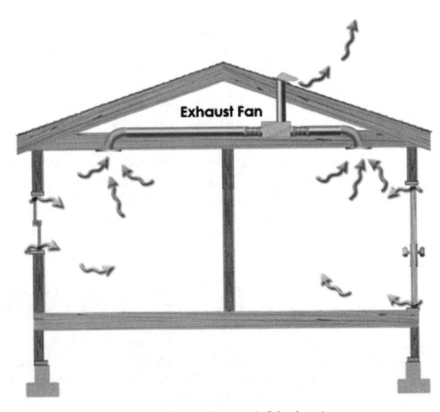

Exhaust Fan

FIGURE 4.1   Exhaust-only ventilation. (Panasonic/Morrissey)

How long can a fan run before it breaks down? The number of hours that a fan can run depends primarily on the lubricants in the bearings of the motor. If a fan motor is rated to run 60,000 hours, it is equivalent to 6.8 years of continuous operation. Although there are many green building programs and residential ventilation codes that require the use of fans "rated for continuous operation," there is no standardized test, group, or agency that actually provides such a rating. It is up to the manufacturer to provide that information.

tively draw fresh air into all the rooms, particularly bedrooms where people spend the most time.

This is the most basic, standard ventilation system. It grew out of the concept of putting a fan in the bathroom to remove odors and a fan in the kitchen to remove cooking smoke and smell. When houses were fairly "leaky," this worked reasonably well, although they could still feel stuffy, particularly in the winter. There is certainly an advantage to using the same fan that exhausts bathroom pollutants to serve as the primary ventilation system. The fan does double duty. (Figure 4.2 is a multi-port exhaust system that removes the air from multiple points simultaneously.)

For this strategy to work, the fan has to run constantly, at least part of every hour. That requires a quiet, energy efficient fan, otherwise it will be defeated, shut off, and stuffed full of socks. And, equally important, the occupant needs to understand and appreciate why it is running all the time or it will be defeated, shut off, and stuffed full of socks.

Since the fan is running continuously and it is quiet, the home occupants will not know that it is on and conversely won't know if it is off or stops running. Why do you need a fan running all the time anyway? Many of the forces of nature are very, very slow. Imagine a stoppered bathroom sink with a leaky faucet, a single drip every 15 seconds. It's just a drop, a tiny amount of water, but after eight hours there will be more than a pint of water in the sink. After a day, it will be approaching a gallon and perhaps the edge of the sink. A small amount of condensation occurring somewhere in a wall will gradually grow mold, encouraged as it will be by that slow, inexorable drip.

Pollutant materials can outgas over a long period of time. The *average* ventilation rate is different than the *effective* ventilation rate. The *effective* ventilation rate, ϵ, is defined as the inverse of the steady-state concentration of the pollutant, the removal of the pollutant from the air in the home.

Consider two identical homes. One is mechanically ventilated continuously at .35 air changes per hour (.35 ACH). The other assumes a low, natural ventilation or infiltration rate of 0.22 ACH for 23 hours a day and one hour of high, mechanical ventilation at 3.4 ACH (see Table 4.1).

TABLE 4.1   Ventilation rate comparison.

System description	Continuous venting at 0.35 ACH, 24 hours a day	Natural 0.22 ACH for 23 hours, and one hour of mechanical ventilation at 3.4 ACH
Average ACH	$\dfrac{0.35 \text{ ACH} \times 24 \text{ hrs}}{24 \text{ hrs}} = 0.35 \text{ ACH}$	$\dfrac{(0.22 \text{ ACH} \times 23 \text{ hrs}) \times (3.4 \text{ ACH} \times 1 \text{ hr})}{24 \text{ hrs}} = 0.35 \text{ ACH}$

FIGURE 4.2   A multi-port ventilator. (American Aldes)

The story goes that a woman's marriage was coming apart. Her husband was associating with a younger model. The woman was moving out of her home in three days. She spent the first day packing her belongings, taking the things that reminded her of better times. The next day the movers came and took the boxes and crates away. On the third day she sat alone at the dining room table by candlelight, put on her favorite music, and feasted on a pound of shrimp, a jar of caviar, and a bottle of a delicate Chardonnay. When she had finished eating, she went into each room and deposited a few half-eaten shrimp shells dipped in caviar into the hollow of the curtain rods. She cleaned up the kitchen and left.

The husband returned with his new girlfriend and all was bliss for the first few days. Then slowly the house began to smell. They tried in vain to find the source of the smell, mopping, and cleaning. The HVAC vents were checked for dead rodents and the carpets were steam cleaned. Air fresheners were hung everywhere. Exterminators were brought in to set off gas canisters. They even had the wool carpeting replaced.

Nothing worked. People stopped coming over to visit. Repairmen refused to work on the house. Finally they couldn't stand it any longer and decided to move out, but they couldn't sell the house. No one would buy it despite the fact that they continually lowered the price. Local realtors refused to return their calls. They borrowed a huge sum of money to buy another house.

The woman called her ex-husband to ask how things were going, and he told her the saga of the smelly house. She listened politely, and said that she missed her old home terribly, and would be willing to take the home back.

Thinking he knew that she had no idea how bad the smell was, he agreed on a price that was about one-tenth of what the house had been worth, but only if she were to sign the papers that very day. She agreed and within the hour his lawyers delivered the paperwork.

A week later the man and his new girlfriend stood smirking as they watched the moving company pack everything to take to their new home . . . including the curtain rods.

The *average* ventilation rate or ACH describes the average amount of air moving through the house. System 1 is a continuously operating, mechanical system with a total ACH of 0.35 including both natural and mechanical ventilation. System 2 relies on a natural, continuous 0.22 ACH for 23 hours along with one, high exhaust hour at 3.4 ACH. The resulting *average* ventilation rate is the same over 24 hours.

The *effective* ventilation rate, ϵ, for the two homes in this example can be defined by considering a pollutant source like rotting shrimp in curtain rods outgassing continuously over a long period of time. For this example the source strength is divided into equal parts of shrimp per hour.

TABLE 4.2   Effective ACH.

	System 1	System 2
Pollutant source strength	1 Shrimp/hour (continuous ventilation)	1 Shrimp/hour (Ventilation at 0.22 ACH for 23 hours then 3.4 ACH for one hour)
Average concentration	$\dfrac{1 \text{ Shrimp/hr}}{0.35 \text{ ACH}} = \dfrac{2.86 \text{ Shrimp}}{\text{volume}}$	$\dfrac{1 \text{ Shrimp/hr}}{0.22 \text{ ACH}} \times \dfrac{23}{24} + \dfrac{1 \text{ Shrimp/hr}}{3.40 \text{ ACH}} \times \dfrac{1}{24} = \dfrac{4.37 \text{ Shrimp}}{\text{volume}}$
Effective (ϵ) ACH	$\dfrac{1 \text{ Shrimp/hr}}{(2.86 \text{ Shrimp/air volume})} = 0.35 \text{ ACH}$	$\dfrac{1 \text{ Shrimp/hr}}{(4.37 \text{ Shrimp/air volume})} = 0.23 \text{ ACH}$

The average daily pollution concentration in the home with the continuous ventilation (System 1) is about half as much as the home with intermittent ventilation (2.86 Shrimp/air volume as opposed to 4.37 Shrimp/air volume). If the pollutant source is continuous throughout the day, continuous ventilation has a greater impact on the *effective* ventilation performance even though the heat loss from the two buildings due to the *average* ventilation rate will be the same.

If the husband and girlfriend had ventilated the house continuously, they would have come home to a much lower level of odor than if they came home and turned a big exhaust fan on for an hour to get rid of the stink that had built up. The stink would still have been there because the shrimp were still there. Removing the source of the pollution is the first step in any indoor air quality control approach. Exhausting air from near the polluting source provides the shortest path from the pollutant to the outside and the least amount of mixing with the room air. Exhaust-only ventilation "sucks" the air out of the house, slightly deflating it. Outside, unconditioned air slips back in through gaps and cracks. If it is hot and humid outside and air conditioned inside, the moisture in the incoming unconditioned air may condense on some surface in the wall or ceiling structure. Although bath and kitchen fans are necessary to extract the immediate pollutants in any climate, exhaust-only ventilation is not the best primary ventilation strategy for hot and humid climates.

SUPPLY-ONLY PRIMARY VENTILATION SYSTEMS

Supply-only ventilation is another primary ventilation strategy (see Figure 4.3). It is not for spot or source ventilation. It wouldn't work well to have a

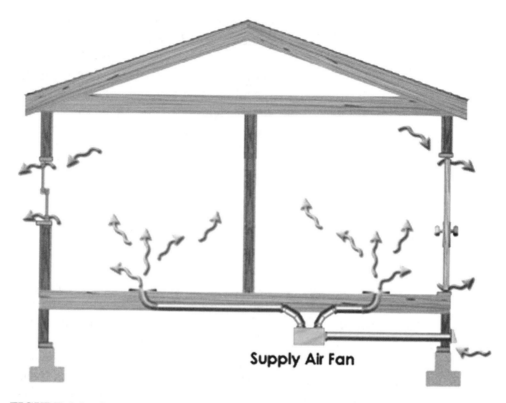

Supply Air Fan

FIGURE 4.3   Supply-only ventilation. (Panasonic/Morrissey)

range hood blowing in, blowing the smoke in the cook's face! Supply-only ventilation works best if it is distributed to the rooms that are occupied for the majority of the time, like bedrooms where people spend about eight hours per day sleeping. A steady, low flow of fresh air to the bedrooms and perhaps other living areas is balanced by leakage from the house and by mechanical airflow through spot exhaust fans.

Because the air is not coming in randomly through cracks and holes in the house, it can be controlled. It can be filtered and tempered, and the rate of flow can be adjusted.

A number of systems are available that supply air directly into the return side of the HVAC air handler. Some of these systems have controllers that monitor the run time of the air handler for heating or cooling and subtract that from the run time required for ventilation. Some of them have motorized dampers that close when outside air is not required. Some have flow restrictors in the inlet pipe that limit the amount of air that is drawn into the system depending on building and system pressures. These are all essentially supply-only systems because they are mechanically adding ventilation air to the building.

In a cooling dominated climate when the air inside the house is predominantly cooler and drier than the air outside the house, a primary supply-only ventilation system will keep the house under slightly positive pressure relative to the outside, keeping unwanted warm, humid air from being drawn into the building system.

BALANCED AND DISTRIBUTED PRIMARY VENTILATION SYSTEMS

Exhaust-only and supply-only systems rely on building leakage to supply the other half of the equation. Negative or positive building pressure balances the mechanical system. Adding a mechanical source of make-up air makes them a "balanced system" (see Figure 4.4). Combining a distributed supply system (bringing filtered air into the bedrooms, for example) with a system that exhausts from the bathrooms could be a balanced system if the two flows are at the same volume and running at the same time. This can be accomplished by using one control or wiring both systems to run continuously.

BALANCED WITH HEAT OR ENERGY RECOVERY PRIMARY VENTILATION SYSTEMS—HRVS AND ERVS

It may seem counterproductive to spend money to condition air and then blow it to the outside. It would be nice if we could blow the pollutants out and keep the heated or chilled air and use it again. That's essentially what HRVs and ERVs are designed to do (see Figure 4.5).

A heat recovery ventilator or HRV uses a heat exchanger to extract the heat or "coolth" from the outgoing air stream and add it to the incoming stream, pre-conditioning it. An HRV or ERV (energy recovery ventilator) is designed and installed in such a way that the supply stream from the outside and exhaust stream from inside the house pass right beside each other (see Figure 4.6). The second law of thermodynamics states, "Heat moves toward cold." So if the supply stream from the outside is cooler than the exhaust stream from the house, the warmth in the exhaust air is allowed to move to the cooler supply air. This has to be accomplished without allowing the two air

Balanced Air-Mechanical In & Out

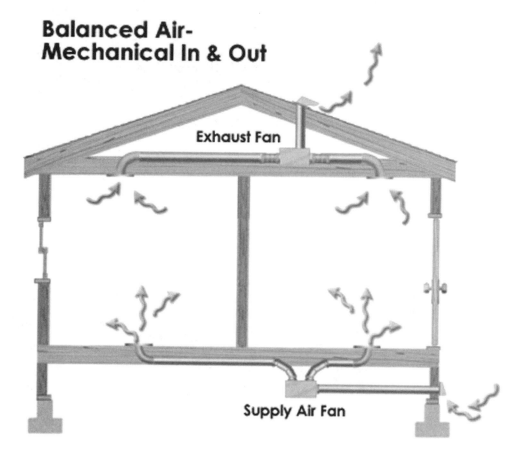

Exhaust Fan

Supply Air Fan

FIGURE 4.4   Balanced ventilation. (Panasonic/Morrissey)

streams to actually mix. If that occurred, of course, the pollutants would just be coming back in and not much ventilation improvement would occur.

This temperature conditioning "magic" occurs by offering a lot of temperature touching surface area in a small box. The most common approach is to use a core with a lot of corrugation-like layers, one oriented one way and the next oriented the other and so forth, building up a cube. Each face of the cube is open to one air stream and closed to the other. The air streams cross each other (cross flow ventilation), touching thermally through all those corrugations but never mixing.

An alternative approach is to use a heat-exchanging wheel that rotates through both streams. A small motor and belt rotate a wheel designed to carry a heat or heat and moisture absorbing medium around and around through the

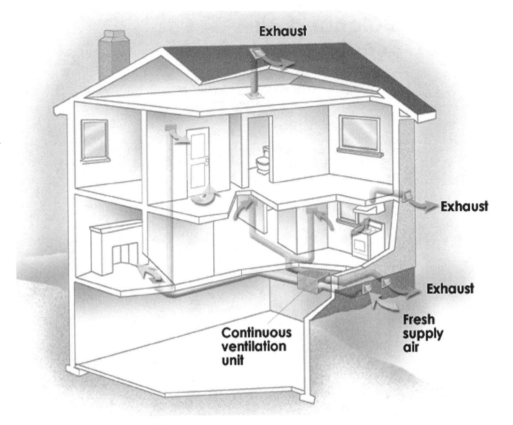

FIGURE 4.5   Heat recovery ventilation. (Panasonic/Morrissey)

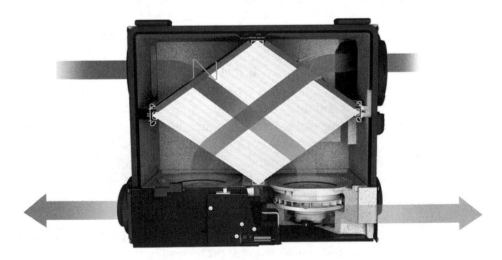

FIGURE 4.6   Heat exchanger airflow. (Venmar)

outgoing and incoming air streams. The wheel absorbs *sensible heat* (heat that can be "sensed" or "felt") as it moves through the warm stream, which is then given up as it moves slowly through the cool stream (see Figure 4.7). If the wheel is filled with a desiccant or moisture absorbing material as well, *latent heat*[1] is also transferred, classifying it as an energy recovery ventilator or ERV. Moisture is captured on the wheel from the air stream with the higher humidity (either due to the wheel being cooler than the air dew point or because of the desiccant fill). As the wheel moves through the drier air stream, moisture is released through evaporation. With desiccant filled wheels, this is a "dry" moisture transfer process since the moisture is in the vapor or gas phase. There are no "wet" surfaces and liquid water does not enter the airstream, limiting the possibility for mold growth. The "fill" for the wheel is typically made of aluminum or, for a total heat recovery, a number of different materials treated with a hygroscopic material such as lithium chloride, alumina, or aluminum oxide, each of which has different moisture-absorbing properties.

Seals or gasketing are included to divide the two moving halves of the wheel, but some air is entrained and carried from one stream to the other which reduces the efficiency. Some transfer occurs because of a difference in pressure between the two sides of the chamber that the wheel is moving through, but the negative impact of this can be minimized by balancing the intake and exhaust fans and designing the housing so that any air leakage that does occur moves into the exhausting air stream, carrying it out of the building.

An ERV does not have to have a rotating wheel (see Figure 4.8). "Flat plate" energy recovery or enthalpy recovery exchangers have a special core material that allows the transfer of water vapor as well as heat between air streams. The moisture passes from the more humid air stream to the drier air stream so when it is more humid outside some of the moisture from the incoming air will be transferred to the drier, outgoing air.

The efficiencies of the heat and energy exchange products have become quite high, some as high as 80 percent. Testing of HVI Certified[2] HRV/ERV products has become extremely thorough, testing the equipment under a range of operating conditions and durations.[3] Moisture and freezing can be a problem in a winter climate. The incoming air is at outside air temperature initially and any moisture that condenses in the fan system can freeze. Certification testing includes cold temperature tests at freezing and below.

---

1. Latent heat is the energy required to change the state of a substance—a gas to a liquid, for example—without changing the temperature.
2. http://www.hvi.org/resourcelibrary/proddirectory.html.
3. See Appendix G for a description of HRV/ERV testing.

FIGURE 4.7 A heat exchanger desiccant wheel. (Venmar)

The difference between an HRV and an ERV is moisture transfer. In a hot, humid climate, the air moving out of an air-conditioned building is cool and dry. The ventilation air coming in from the outside is warm and humid. Bringing more humidity into the building will increase the work for the air conditioning system to dry the air out. If some of that incoming humidity can be carried out in the exhaust stream, there is an energy savings—a good application for an ERV.

FIGURE 4.8 A single room ERV. (Panasonic)

There are a number of different efficiency numbers listed in HVI's Certified Products Directory for HRVs and ERVs. All the different information can make it very confusing to choose between systems. The number to look at is the sensible recovery effectiveness (SRE) that defines the "sensible energy recovered minus the supply fan energy and preheat coil energy, divided by the sensible energy exhausted plus the exhaust fan energy. This calculation corrects for the effects of cross-leakage, purchased energy for fan and controls, as well as defrost systems."[4] See Chapter 9 and Appendix G for more information.

4. HVI, "Heat Recovery Ventilators and Energy Recovery Ventilators," http://www.hvi.org/resourcelibrary/proddirectory.html.

Climate maps don't show up well in black and white, but there are a number of them available on the Internet that provide indepth information regarding local climates.[5]

In a heating climate, the incoming air stream may be cold and dry. Transferring some of the moisture from the exhaust or outgoing stream will warm and humidify the incoming air and make it more comfortable. But if the house is tightly constructed, it may be better to expel the moisture rather than trapping it in the house—a good application for an HRV.

MATCHING THE SYSTEM TO THE CLIMATE

Clearly one type of central ventilation system doesn't fit all applications or all climates (see Figure 4.9). As a starting point, however, in order to keep humidity from being sucked into the building system, positive pressure ventilation works best for a climate that relies on air conditioning for much of the year. Negative pressure (exhaust only) ventilation works well in a climate that is dominated by the heating system. Mechanically balanced ventilation works well in both cases, and again in terms of a starting point, an ERV limits in the infusion of humidity in a cooling climate, while an HRV may be better suited to a heating climate (see Table 4.3).

CIRCULATION AND DISTRIBUTION BASICS

It is certainly a good idea to remove the pollutants near the source with spot ventilation systems like bath fans and range hoods. The make-up air for these devices leaks in randomly around the house through gaps and air leaks, windows and doors, down chimneys and through other exhaust vents that are not being used. Those gaps, leaks, and holes are randomly located throughout the house. The air may be coming from the basement or crawl space, the garage, or the furnace closet, carrying with it pollutants of all sorts. When insulation in the attic or in the wall is removed and it is black or there are black streaks on the carpeting or the baseboards, that's the filtering system for incoming air. One of the best things about an engineered ventilation system is that the

5. For NOAA climate maps go to http://cdo.ncdc.noaa.gov/cgi-bin/climaps/climaps.pl.

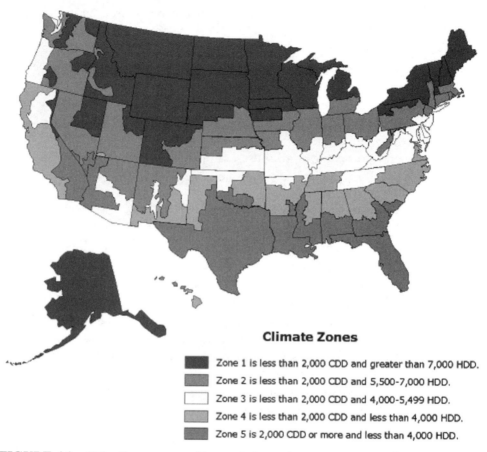

FIGURE 4.9 U.S. climate zones. (Energy Information Administration)[6]

TABLE 4.3 System choices for different climate conditions.

Air Conditioning Dominated Cimate Hot & Humid	Heating Dominated Climate Cold
Positive pressure—supply-only (dehumidifying supply)	Negative pressure—exhaust-only
Mechanically balanced—ERV	Mechanically balanced—HRV
Spot ventilation baths	Spot ventilation baths
Spot ventilation kitchen	Spot ventilation kitchen
Spot ventilation—garage exhaust	Spot ventilation—garage exhaust
Radon/soil gas mitigation system	Radon/soil gas mitigation system
Attic exhaust	Attic exhaust

6. http://www.eia.doe.gov/emeu/cbecs/climate_zones.html.

source of the air is known and it can be filtered and conditioned before it is supplied to the house.

Residential ventilation system design is more complicated than office ventilation system design where adults spend most of the day sitting. In an office, fresh air can be introduced at the four foot level and cover most situations. In a home, babies crawl on the floor, people sit at tables and desks and on sofas and lounge chairs and lie in beds, people stand in the kitchen and walk through the halls. Ventilation air should be mixed, diluted, filtered and delivered to people at all these heights. Fresh air should be distributed throughout the home, which can be done in several ways.

Distribution with a Central Air Handler

If the house is conditioned with heated or chilled air, it already has a distribution system. Unfortunately, most of the time the HVAC (heating, ventilating, and air conditioning) contractors ignore the "V" part. The conditioned air ducting is designed to circulate warmed or chilled air to each room and then draw the air back to the air handler to re-warm or re-chill and dehumidify it. Ideally there will be a supply and a return in each room so that the system will work effectively whether or not the door to the room is open or closed. Commonly, the small number of central returns will be located in the hallways and the system will rely on door under-cuts, transfer grilles, or "jumperducts" to allow the air to flow from the rooms back to the return. (See Chapter 8 for more information on transfer grilles.)

The circulation system should be designed for the climate in which the house is located using computational methods called Manual J and Manual D[7] developed by the Air Conditioning Contractors of America (ACCA). These approaches analyze the structure of the building to determine how much heat will be gained or lost in each room and match the delivery system to compensate so that the room will be comfortable throughout the year. Since in virtually all the climates in the U.S. a house goes through both heating and cooling seasons to some degree and the HVAC system needs to be able to handle both, the design is a compromise in terms of grille locations and sizing.

The grilles in a room, for example, need to be able to "throw" the air far enough out into the room so that all the air in the room is conditioned and comfortable. But if the grille is too low and the air is chilled, it will cause a

---

7. Manual J is used to define the heat loads of a house on a room-by-room basis.
   Manual D is a process for designing the duct layout that will satisfy the design conditions defined by Manual J most efficiently. http://www.acca.org/design.

"draft" and be uncomfortable. The "throw" depends on the location of the grille, the velocity of the air coming out of the grille, and the design of the grille itself. (Smaller grille openings increase the velocity of the air leaving the grille and "throw" it farther out into the room. Smaller grille openings can also reduce the amount or volume of airflow and increase the noise of the air moving through the grille.)

In a heating climate, the supply grilles are often located under the windows on the outside walls so that the rising, warmed air provides an "insulation blanket" in front of the glass to keep people warmer. That location will not work as well for chilled air. Carpets or furniture often block grilles that are located in the floor or low on the wall. It is important for the designer to think about the complete path of the air from the grille, through the room, and back to the return. Appreciating how the room will be used is critical for having a functional system that satisfies both the engineers and the occupants.

Grille manufacturers use computer simulation tools called computational fluid dynamics (CFD) to see how air moves through the room. Quite often the location of the grille or the face velocity can be too weak to allow the air to sweep through the whole room and back to the return, leaving dead air spots where there is inadequate circulation.

If there is no return in the room, then the air coming in will pressurize the space. It will seek to leak out through gaps and openings and hopefully under the room door when it is closed back to the return grille. If the house has a central return system rather than returns in each room, the rooms must be considered as ducts and the room doors as dampers. If all the air that is supplied to the room can't get back to the return, then the space where the return grille is located will be at a slightly negative pressure as the system tries in vain to suck the air back to the air handler. This slight amount of negative pressure in an air conditioning dominated climate will draw humid air into the building system from the outside through the cracks and leaks just as an exhaust-only ventilation system would.

It is worth emphasizing that the HVAC system is designed for heating and cooling and not for ventilation. Locating supplies on the floor under windows for example is not ideal for introducing cool, winter ventilation air into the room. The air may just lie on the floor and never reach standing or sitting people. If the grilles are located near the ceiling in a cooling climate to optimize air conditioning, the warmer outside air may never drop to people height. Grilles for the HVAC system may be too large to offer adequate "throw" for the lower velocity ventilation air. The ducts may be too big for the smaller power ventilation fans and the air may just "drop" out of the grilles closest to the system and not move through the whole system. The ventilation air may never reach the rooms at the end of the line. The system designer should consider all these issues when combining the ventilation system with the heating and cooling system.

Many of these problems can be avoided if the two systems run simultaneously, but that means the electrical penalty of running both fans. If high efficiency furnace fans (with ECM or electronically commutated or controlled motors) are used, the operating cost will be reduced and the penalty minimized.

One other consideration in a cooling climate is that moisture condenses on the cold surface of the cooling coil, drips into the condensation pan, and runs out the drain, drying the air in the process. If the fan is run continuously to improve ventilation, the moisture on the coil can re-evaporate into the air stream and be delivered back to the house, reducing the cooling effectiveness.

Adding a Fresh Air Inlet to the Air Handler

The beauty of adding ventilation to the fully ducted HVAC system is that the full distribution system is there. Adding the "V," the ventilation, can be relatively easy. Fresh air will be distributed throughout the house, and limiting the number of supply points from the outside limits the difficulties of providing filtered and tempered air.

A duct to the outside can be added to the return side of the air handler. Outside air drawn in through that duct when the air handler is running will be mixed with the air from the house, conditioned, and delivered back to the house. The air handler runs in response to the house thermostat, turning on to heat or cool the house. If the weather is mild or if the system has been set back for night or "away" mode, the air handler might not turn on for hours, and consequently, there will be no ventilation. Night set-back is great for saving energy but not for circulating ventilation air.

To compensate for this, controls have been developed that will monitor the run time of the air handler and compare it to a preset ventilation schedule. If the air handler has not run long enough to meet the schedule, the control will activate the blower again for ventilation. Some of these controls work with motorized dampers, leaving the intake open only when the blower is running.

The location of the air inlet is important. This is effectively the "nose" of the house. It shouldn't be near bad air sources like behind the garbage cans or in the carport. It needs to be kept clean. If there is a filter on the intake, it must be accessible for servicing.

Some of them have temperature and humidity sensors to override the ventilation schedule if the outside air is too hot, too cold, or too humid. These overrides keep the load on the heating and cooling system down, but it is a trade-off with moving fresh air through into the house.

This approach does put the home under slightly positive pressure, relying on leakage and spot ventilation fans to balance out the fresh air that is being drawn into the air handler.

Adding an HRV or ERV to the Air Handler

Another balanced approach is to tap an HRV or ERV into the HVAC ducting. This approach saves the cost of a separate ducting system for the ventilation system. The HVAC system should be balanced with the same amount of air moving out of its supply registers as is coming back in through its return grilles. It should be designed to be a closed loop circulation system, moving the air from the inside through the conditioning system and back to the inside.

An HRV or ERV system is also a closed loop, with the same amount of air being drawn into the house as is being expelled from the house, moving the air from the outside to the inside and back to the outside. If either of these paths is poorly installed, integrating them will affect both of them. Both HRVs and ERVs are designed to be self-contained with no impact on other devices such as fireplaces, gas water heaters, clothes dryers, or range hoods.

Air from the outside is drawn into the HRV or ERV, moves through the heat/energy exchanger core, is supplied to the return side of the air handler, and is distributed throughout the house. Air from the house is either drawn from the return registers or from the polluting points like bathrooms, then moved through the HRV or ERV, and expelled to the outside.

Careful design of these systems is important to limit the impact of unconditioned outside air on the conditioning system. HVAC system manufacturers have specifications on the temperature of the air that directly impacts the conditioning coils. Hitting a heating coil with freezing air can reduce the life of the coil. Even in systems with an attached ERV with latent (moisture) energy recovery, the dew point of the air entering the system can be much higher than the air in the chilled supply ducts, resulting in moisture issues and mold when the ERV is running but the central fan isn't. By inserting the supply air on the return side of the air handler far enough upstream, the air has a chance to blend with the house air and will impact the chilled, supply ducts at an acceptable temperature and dew point.

Although this approach is frequently used, there are a number of drawbacks. It is difficult to balance the HRV/ERV because of the ducting and operating variations of the HVAC system. The exhaust side of the HRV/ERV works against the negative pressure in the return plenum as it tries to draw air out of the air stream returning to the air handler. If the negative pressure is sufficiently low, it can severely reduce or even stall the exhaust airflow through the HRV/ERV.

Since the distribution approach is designed for the HRV/ERV to run only when the air handler is running, mechanical ventilation may not be supplied when it is most needed—in the spring and fall when the temperatures are moderate enough to not require heating or cooling. Ventilation rates may be excessive during the other seasons when the air handler is running frequently.

If the HRV is coupled to a central system in a cooling dominated climate, the outside air, which will be full of humidity, may cause condensation on the interior surfaces of the HVAC equipment and the supply plenum and ducting. It would not be a positive function for the primary ventilation system to be the source of indoor mold!

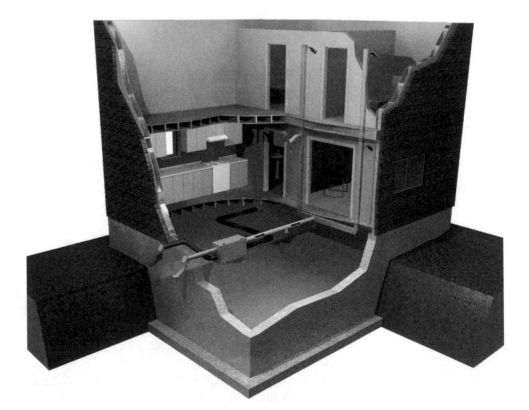

FIGURE 4.10   A balanced HRV system. (Venmar)

Distribution without a Central Air Handler

Since the HVAC system is designed to be a closed and balanced circulation system—taking air from the house, conditioning it, and returning it to the house—it is a bit of a compromise to add ventilation air to the system. It certainly can be considered another form of conditioning of the air like filtering or heating or cooling or humidifying/dehumidifying it (see Figure 4.10). But, as can be seen from the above discussions, it can also cause issues with humidity and impact on the mechanical equipment. There are many homes that are heated without ducting, using hot water or electric heat, for example, that require stand-alone ventilation systems. Ideally such a system is fully ducted, purposely and mechanically removing stale air and delivering fresh air to each room and not relying on the rooms, the doorways, and the halls to act as the ducting of the system. Such a system, however, never happens. It's a ventilation system engineer's dream!

"Fully Ducted" Primary Ventilation

What stands as a "fully ducted" system has both a mechanical supply and exhaust and uses the rooms and halls as ducts to convey the air between the two (see Figure 4.11). It is critical that there is a means to bypass closed doors for this to work either with sufficient door undercuts, jumper ducts, transfer grilles, or what some building codes call "return air pathways." A small supply system draws air in from a single outside point providing the option for pre-conditioning and filtering and delivers the air to the spaces where people spend most of their time, like the bedrooms and living room. The air moves through the rooms toward the exhaust points located in bathrooms or kitchen. The supply system may use small 2 inch or 3 inch ducting, allowing it to be simply run through the wall system. The volume of air to each supply point is low, less than 20 cubic feet per minute, at a velocity that should be barely noticeable in the room.

One important note is to recognize the difference between the delivery system for heating and cooling and the delivery system for fresh air ventilation. Air is not very good at holding heat. One cubic foot of air can only hold .018 Btu per degree Fahrenheit rise of temperature. To change the temperature of the air in a room, you need to add a lot of "new" air at a higher or lower temperature. That's why HVAC systems use big fans and big ducts. For ventilation, however, a small amount of fresh air has a big effect. A 7500 square foot house with seven bedrooms only needs 135 cubic feet per minute of ventilation air (according to ASHRAE). A 3000 square foot three bedroom house only needs 60 cfm. The ducting and grilles for a ventilation system can be much smaller since the air will be blending with room air at a low velocity. If

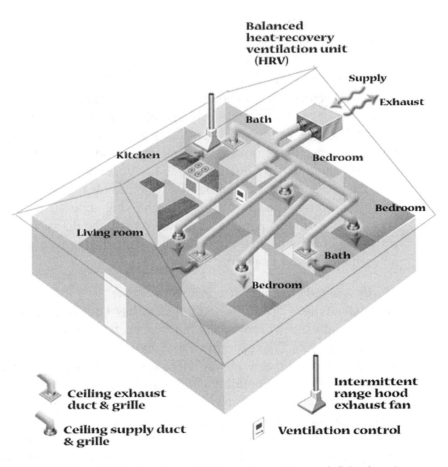

FIGURE 4.11 A fully ducted ventilation system. (Panasonic/Morrissey)

the grilles are properly designed so that they create a minimal face velocity, just enough to "throw" the air out gently into the room, and installed so that they are out of the human comfort zone, the flow from them should be hardly noticeable.

Sound level also must be considered as with any air moving system. Smaller grilles and smaller ducts can amplify air noise. The system can use a highly effective filter (MERV 11 rating or higher), but it is difficult to get people to maintain the filters on their primary HVAC system. Maintaining the filter on any ventilation system is perhaps even more critical since it is designed to provide the fresh air that the home's occupants need to breathe and stay healthy. For maintenance reasons, the fan/conditioning/filter housing should be located where it is accessible and relatively obvious.

Exhaust with Passive Inlets

Using constantly running exhaust fans or a central ventilation fan can put the house under slightly negative pressure. "Passive" inlets or "smart holes" or "trickle ventilators" can then be used to allow make-up air back into the house. (More on the design of these devices in Chapter 8.) For these systems to work effectively, the house needs to be tightly constructed, tightly enough so that a low volume exhaust fan will able to pull it to a negative pressure no matter where the inlets are located. Because of the "stack effect" in the house (the rising warm air) and because of the wind loads pushing on different sides of the house at different times, and because of leaks in the HVAC system ducting, the pressures in the house are unpredictable. A passive inlet located in the wall of a second floor bedroom may actually serve as an outlet.

If the inlet is working properly, however, it will provide a simple, non-ducted means for delivering fresh air where it is needed, particularly to bedrooms. Air moves through the room, through the doorway, into the common space where it is vented by the exhaust fan. Note that if the inlets and the majority of the air is coming into the space from the first floor and exiting through an exhaust fan located on the second floor, it is likely to bypass rooms on the second floor, especially if the doors are closed. As simple as this design appears to be, it relies on a thorough knowledge of building pressures by the designer, the installer, and the building occupants to work effectively.

HRVs and ERVs

HRVs and ERVs are designed as stand-alone systems and consequently they can be used easily <u>without</u> connecting them to a heating or cooling system. They bring in outside air, condition it, deliver it through dedicated ducting, extract stale or "used" air, and exhaust it to the outside. Like all of these systems, the ducting needs to be well designed and installed so that the system is balanced. There is a level of tolerance in the ducting for these systems, however, because the heat exchanger core and filters often have a much higher resistance to airflow than the ducting. That means that the performance of the system can be effectively pre-engineered simplifying the installer's job (although it does not completely remove the onsite design and balancing that must be done as well).

Ventilation Distribution Summary

Ventilation air is a critical factor in a home. It is certainly as important as the color of the walls or the tile on the bathroom floor and must be considered as a fundamental issue right from the beginning. The problem is that people can't see what's in the air. They can't see the formaldehyde or the carbon monoxide, and they might not even attribute how they feel in their home to the quality of the air. They know when it is too hot or too cold. They can even feel when it is too dry or too humid. But they don't know when the air quality is bad.

Ideally the primary ventilation system would stand on its own with a dedicated ducting, circulation, and distribution system. Anything other than that—integrating it with the HVAC system or using the rooms as ducts—is a compromise. National standards like ASHRAE 62.2-2007 offer an averaged, national place to start in choosing the ventilation system. Some states and codes mandate local systems, and many green building programs require ventilation as part of their criteria.

There are many options for ventilation system design. No one design will fit all applications.

Chapter 5
SOUND

Just like the distribution of ventilation air, in the past the sound or noise of the air moving device wasn't a major factor in selecting the fan. Bathroom exhaust fans have always been used with the intention of removing excess moisture from the bathroom, even if it was only to control mirror fogging. However, many consumers and even some builders considered the bathroom fan's purpose to be odor control. People joked that bath fan noise was a benefit because it covered up the rude noises in the bathroom. With tighter houses and fans being used as the primary ventilation system required to run all the time, reducing the sound level of the fan became a serious concern. As houses become tighter, outside background noises are hushed. Internal noises become much more noticeable—the drip in the sink, the running toilet, the refrigerator, the HVAC system running in the basement, or the bathroom fan running for ventilation. The occupant of the home notices these things, and if they don't have a good understanding as to why they are running, will seek ways to shut them off. The interrelationship between a quieter home and the need for a quieter ventilation system emphasizes the obvious systems nature of a building and particularly a house. The house (and its occupants) are a system.

Unfortunately you can't listen to an installed fan before you buy it, and the contractor that installs it doesn't have to live with it. Therefore sound ratings must be numeric values that enable the person making the selection to rank products in the same order that average people would rank them if they simply heard them running. That means the ratings must be comparable. It also means that the ratings must be in terms of the human response to sound. And lastly, the sound test must be repeatable. If a fan is tested several times at long intervals of time, the measurements must be the same within reasonable tolerances.

GENERAL SOUND CONSIDERATIONS

The most common engineering approach to describe a sound level is to talk about "decibels." But to the average mortal, decibels don't make a lot of sense. The decibel scale is logarithmic. It rises on an increasingly steep curve.

Decibels are the ratio of a sound unit to a base unit, expressed logarithmically; a decibel is an exponent of 10. Twenty decibels is 10^{20} or 100,000,000,000,000,000,000 times louder than the base unit. Something that is 100 decibels is not twice as loud as something that is 50 decibels. It is much, much, much louder than that. Fifty decibels might be a couple having a quiet conversation. A hundred decibels is the sound of a jackhammer or a noisy rock concert. It is just not intuitive that 43 decibels is twice as loud as 41 decibels, for example. The useful thing about the decibel scale is that it is relatively simple to measure, and it makes it possible to work with fewer zeros (see Figure 5.1).

Many activities are measured in decibels and, even for sound, there are different kinds of decibels, e.g., dB(A), which means the decibels are weighted for some type of response to the character of the sound. (In the "A weighting" scale, the sound pressure levels are "weighted" to more accurately reflect the human ear by reducing the lower frequency and higher frequency bands before they are combined to give a single sound pressure level value.)

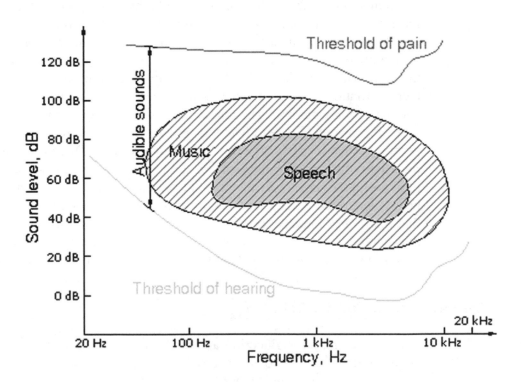

FIGURE 5.1   The measurement of audible sounds. (dbx Consulting)

To sell quiet fans and to be able to compare one to the other, a more understandable means for measuring sound needed to be developed. A metric was needed that increased in a linear manner. Along came the *sone*.

According to the Oxford English Dictionary, one of the first uses of the word sone in English literature was from a poem in 1616, where the poet (J. Lane) was waxing eloquently about the "music of the spheres" and needed something to rhyme with "tone." Much later, in 1936, Dr. S. Smith Stevens, while he was head of the Harvard Psychoacoustic Laboratory, proposed the sone scale as a linear indication of perceived loudness. People, unfortunately, perceive noise or loudness in different ways that are difficult to quantify. By measuring the decibel level at different frequencies or pitches and combining them, a linear scale for sones was developed that is simpler to understand. A fan that is rated at 4 sones is twice as loud as a fan that is rated at 2 sones.

The problem is that without some sophisticated equipment, measuring sones is much more complex than measuring decibels and converting one to the other is, to say the least, difficult. Rather than pulling out a relatively inexpensive decibel meter and pointing it at a fan, measurement has to be done in a sophisticated sound chamber with a clearly prescribed location for the microphone and a lot of other elements. Human beings are pretty good at pinpointing a sound and mentally filtering out all the background components and sound reflections in the room, etc., but what you hear and what your neighbor hears are very different, not something you can put on the end of bath fan packaging as an official rating. To get a repeatable and comparable sound or sone rating the product needs to be tested in a laboratory that has been constructed to standardized criteria such as the Texas A&M Environmental Systems Laboratory[1] in College Station, Texas, which is where the Home Ventilating Institute (HVI) tests and certifies home ventilating products. (Figure 5.2 shows the door to ESL sound chamber.)

This measurement process becomes even more difficult as fan products get quieter and quieter. When the sound level is down to one half of a sone, the smallest of background noises becomes ever more critical. Even in the carefully constructed acoustic chamber (where you can hear your heart beating) at the remote location of the Energy Systems Laboratory (ESL) of Texas A&M, they have to wait until late at night when trucks stop passing on the highway, 3 miles away, before they can get accurate information!

After the fan is tested for airflow, it is installed in the "diffuse, reverberation chamber" at the ESL lab without making any modifications to the prod-

1. http://esl.eslwin.tamu.edu/riverside-lab.html.

FIGURE 5.2   A sound chamber. (ESL)

uct. The chamber is constructed with heavy, multi-layered, insulated, non-parallel, reasonably airtight walls with hard (reverberant) interior surfaces. It is approximately 25 feet long, 20 feet wide, and 12 feet high. It is supplied with make-up air through an insulated, labyrinthine inlet duct with a throttling device to control the rate of flow. The outlet from the chamber is through a rectangular isolation duct, into an anechoic muffler that opens out into the atmosphere, avoiding the entry of environmental sound. The chamber is built on a resilient base that minimizes the transmission of ground-borne (sound) vibration. The single personnel opening is equipped with two heavy, insulated doors with seals and positive latches. A microphone is mounted in the chamber on a rotating boom that is capable of moving the microphone through approximately 180 degrees during the 30 second data collection period. As fans get even quieter, the sound of the motor rotating the microphone has become the greater noise source and is being replaced with a series of fixed-point microphones. (Figure 5.3 shows the rotating microphone in the sound chamber.)

The only meaningful basis for rating a fan's sound is its *sound power*, but there is no single instrument for measuring that. The rotating microphone or microphone array measures *sound pressure* on their diaphragms.

Four sound pressure measurements are conducted in 24, one-third-octave bands. The first measures the fan and the background noise so that the background noise can be subtracted. The second measures just the background noise. The third test measures the background noise plus the sound from a precisely calibrated reference sound source (RSS) without the test fan running. (The RSS is an expensive noise generator that is regularly sent to a standards laboratory for calibration.) And the fourth measures the test unit, background noise, and reference sound source running simultaneously to check for background steadiness.

Every sound measurement chamber has its own peculiar characteristics. By running both the fan and the RSS in the same chamber in quick succession, the two can be compared without regard to the room's character because the room will behave the same for both devices. The sound power (or noise level) of the test unit is determined by mathematically comparing sound pressure measurements in the chamber to sound pressure measurements of the reference sound source. For this to be consistent, the chamber must meet specific requirements especially related to standing waves and background quality.

The sound power the fan is producing in the 24 bands produces 24 sound power numbers that are not related to human perception. Converting those numbers to human perception requires two steps.

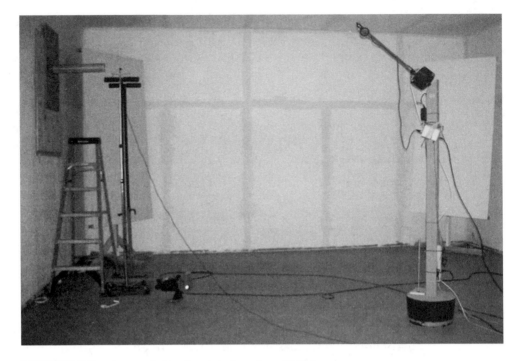

FIGURE 5.3   A sound chamber interior view. (ESL)

The first step is to recognize the fact that humans do not perceive various frequencies in the same way. The human auditory system is most sensitive in the middle ranges and less so at the lower and higher frequencies. There are also differences between people. Dr. Stevens measured a vast number of research subjects comparatively. He started in the frequencies where people have great sensitivity, 1000 Hz at 40dB, and asked people to tell him when another frequency was equally loud. He also asked when two tones were twice or half each other. This provided a reliable set of data representing the response of normal humans to a variety of frequencies and sound levels. These data were next reduced to mathematical curves and applied to all 24 of the frequency bands. The result is the "equal loudness indices" table where the computer looks up each single-band sone number.

The second step is two steps in one. The 24 bands are combined for total loudness, and the human response to dominant tones is factored into the equation used for totaling. When a sound spectrum has dominant tones, it is perceived as being louder than a "white noise" spectrum that an instrument might measure as being the same. In other words, the 24 bands are simply added, and

TABLE 5.1 Relative sound levels.

Source	Sone level
Low cost range hood	7.5
Low cost, noise masking bath fan	4.5
TV set at normal living room level	4
Max level for intermittently operating ventilation fans in ASHRAE 62.2-2007	3
Washington State Ventilation Code bath fan sound level	1.5
Continuously operating ventilation fans in ASHRAE 62.2-2007	1
Quiet refrigerator in a quiet kitchen	1
Very quiet computer fan	.6
Really quiet bath fan	.3

the dominant band is added after being reduced by the penalty for dominance.

Since the sone is a measurement of how the human ear perceives "noise" and since our ears hear noise differently at different frequencies, the sone measuring process "listens" to the 24 frequency bands and then assembles all that information to arrive at the certified single fan loudness number weighted for human dominant tone sensitivity in sones[2] for the fan under test.

It's strange that people like to ride around in their cars with the stereo blasting and the entire car shaking, but they complain about bath fan noise. There are so many annoying abrupt sources of sound in our world like cell phones and motorcycles (and even the impact of titanium golf clubs!). Tightly constructed, quiet homes are havens where the occupants can control the environment to their heart's desire.

SOUND CONSIDERATIONS FOR THE PRIMARY VENTILATION SYSTEM

The primary ventilation system should run constantly, changing the air throughout the day, and if not acoustically invisible, the homeowner will defeat it. If the product used for this purpose is located in the room, it should operate at 1 sone or less. If it is remotely mounted, the installed system should operate in the same way, effectively at 1 sone or less. To test this, the refrigerator might have to be unplugged to hear the fan running or it may mask the sound.

The simplest way to achieve a 1-sone level, if the fan is installed in the ceiling, is to use a fan that is rated at 1 sone or less. To achieve the 1 sone

2. HVI Publication 915, "HVI Loudness and Rating Procedure."

operating level once the fan is installed, it is important to follow the manufacturer's installation instructions. Sounds generated by the fan motor can be transferred to the building structure through the mounting. Building surfaces can act as large sounding boards that will amplify minor fan noises.

If the fan is capable of moving more air than is required for continuous operation, the fan speed can be reduced with a speed control if the fan is rated for variable speed. The method used for speed control sometimes induces a "hum" in the fan motor that is disturbing. This may be from a single operating point in the control, and can be reduced by adjusting the fan speed above or below that point to remove or reduce the "hum." This may also be the case with a remote mounted fan.

Occasionally with a remote mounted fan the air moving through the ceiling grille can produce objectionable noises. Changing the grille to a different configuration may help to solve or reduce the problem.

Mounting is also important for HRV/ERV primary ventilation systems. Their motor noise can also be transferred to the building structure and amplified. Some HRV/ERV manufacturers supply mounting straps that "hang" the unit away from building components to isolate operational hum. (Remote mounted fans (in-line fans) and HRV/ERVs are not rated by HVI for sound level because they are mounted remotely from the living space and every installation will be different making a prescribed, laboratory mounting configuration unrepresentative.)

In some cases the fan will be too quiet and the occupants won't even know if it is on or off. If the air quality in the home and the health of the occupants is dependent on that fan running all the time, it is extremely important that it keep running. It is vital that the occupants know what it is and what it is doing, and it may be advisable to add an indicator light, labeling, or fan-proving indicator that provides a clear indication that the system is working. If the house changes hands, a new occupant may have no idea why there is an exhaust fan running all the time.

SOUND CONSIDERATIONS FOR BATH FANS

There are times when a bath fan that created some background noise would be advantageous. Such fans are available and are not well suited for primary ventilation systems. They make too much noise. Turn 'em on. Get the job done. Turn 'em off. Quite often little or no thought was put into the installation of these fans. Just because they are making noise doesn't mean they are moving any air. Try holding a piece of tissue paper up to the grille and see if the air draws it up. Noise does not equal air motion!

In order to minimize cost and size, the noisier fans generally use a small "blower" wheel and spin it fast. The air squeezes in through the grille and is pushed out through a small, often a 3-inch diameter, duct. All of those components make the fan noisy. It is like a lot of water moving through a river and reaching the rapids. Banging around the rocks and through the narrow passages, tumbling and falling, it makes a lot of noise. When it reaches the unencumbered, wide open stretch it quiets down and flows smoothly.

To quiet the fan, manufacturers began to make the system more efficient, removing the sharp edges, widening the openings, using a larger blower wheel and spinning it more slowly. All of the components of the products contribute to their efficiency and performance. The fan is a system. The number of blades on the wheel and the size of the air inlet openings in the grille contribute to the efficiency, as well as the noise the fan makes. Fan designers are trying to make products attractive, both aesthetically as well as economically.

Some fan motors will develop a "hum" when operated with a speed control. The full AC power that comes directly to the fan is a sine wave, alternating smoothly from full positive to full negative and back again, 60 times per second in the U.S. (60 cycles or 60 Hz). Solid-state speed controls "chop" that sine wave, using an electronic switch like a transistor to turn on and off, providing part of each wave to the fan motor. The more the switch is off, the slower the fan runs. It is sort of like switching the room lights on and off and on and off, slowly or rapidly. It will make it dimmer in the room than just leaving them on. There are some versions of these devices that are more sophisticated, using microprocessors to carefully adjust the fan speed for minimum motor impact and hum generation.

There are other fan speed control approaches such as *variable reactance transformers* that will provide the full sine wave at lower amplitude. Some motor speeds can be adjusted by using matching capacitors for different operating points, and some motors are specifically made to operate at different speeds and have a number of "windings" on their motors. These are generally the bigger motors like the furnace blower motors or some of the large attic fans. There are also "electronically commutated" or EC motors that are specifically designed to be used with their own matching speed control. These motors use only the power necessary to perform the required task and almost always include the control. As of this writing, however, these motors are rarely found in residential ventilation products because of cost, but that is changing.

Make sure that any speed control you use has been safety tested by an agency like UL, ETL, or CSA or a similar laboratory, and make sure that the fan manufacturer has approved its use since the device may overheat the motor or burn it out in a short time.

And that forces compromise. When the sound level of the product gets below one half sone, however, everything has been done to keep the fan quiet. As we age, even the 1.5-sone fans are a stretch to hear!

If an existing bathroom has a noisy, annoying fan to start with, there isn't a great deal that can be done in terms of the installation to make it quieter. Often it will be poorly installed, making the noise worse. If the ducting is kinked and twisted and has numerous sharp, 90 degree turns, increasing the diameter of the ducting and straightening it out will make the air flow more easily and reduce a small amount of the fan noise. Replacing small diameter ducting with a larger diameter will also quiet the fan, making it easier for the air to move through the system and out of the building.

Replacing a noisy fan with a quieter one is the best way to keep the system noise down. A straight, smooth, reasonably large diameter duct will certainly help. Following the manufacturer's installation instructions helps a great deal.

Another alternative is to use a remotely mounted fan or an in-line fan. By removing the fan from the room, most of the noise is also removed. Kits are available to retrofit a remote mounted, in-line fan to an existing ceiling mounted fan box. Disconnecting or removing the existing fan wheel and motor will ease the flow of air through the housing, improve the operating efficiency, and keep the noise level down. Most of the in-line fans can be operated with a speed controller. (Check the product manual. If it doesn't say "Not for use with a solid state speed control," then it probably is okay.) The advantage there is that the fan can be slowed down. If it is operating slower, the air will be moving more slowly and generally making less noise. The speed of the fan can be manually adjusted for an acceptable sound and air-flow level.

Besides adjusting the speed, there are other new technologies coming along that will quiet the operating noise of the fan. Manufacturing a fan to be really quiet requires compromises sometimes in the "strength" of the fan: its ability to aggressively push or pull the air through the ducting. Ventilation has become so important to healthy buildings that manufacturers are working hard to develop new technologies and approaches and provide better and more versatile products.

For a new installation, all the options are open. Starting with a quiet product, designing and employing a good ducting design (at least as good as the manufacturer's instructions), making it easy for the air to move smoothly from the inside of the house to the outside, will all make the system quieter.

Note that manufacturers do want these products to operate as they conceived them. They test them. They poke them. They prod them. They are required to put a great deal of information in product manuals warning about health and safety, but there is also a great deal of good information about how to install them so they work correctly. Failing to read the product manual is a common mistake. After all, how hard could it be to put in a bath fan? But like everything else about the product, from the backdraft damper to the packaging, the product is a system. By ignoring the installation manual, the product is not likely to work as described.

SOUND CONSIDERATIONS FOR KITCHEN FANS

A kitchen range hood is designed to remove the airborne cooking effluents, carrying the smoke, grease, and odors out of the working space. In a commercial application, that process is required by code to be at a high enough velocity to prevent the grease particles from settling where they might cause a fire hazard. Grease spatters and droplets fall rapidly downward without being captured. Grease leaves the cooking zone as a vapor and cools quickly into very small (aerosol) particles that remain airborne for hours. It takes a lot of air movement to remove them and a lot of air movement generally means a lot of noise, especially if the fan or blower is located right there in front of you in the range hood.

The fan in the range hood has to be powerful enough to move a large amount of air from the area around the stove, draw it through grease filters, and push it through a length of ductwork to exit the building through a vent or hood. Range hoods are available that are rated at 1.5 sones, although their flow rates are relatively low.

Mounting the fan remotely, such as on the exterior of the building, helps, and you will almost always see commercial exhaust fans remotely mounted on the roof or sidewall of restaurants. This is both for noise and for size constraints. Since the ducting for a range hood needs to be smooth and made of metal, the sound will echo back into the room. Sound mufflers that work like a car muffler are available to install in the exhaust duct.

Most range hoods have speed controls as well, although slowing them down will obviously reduce their airflow as well as their sound level.

If it is an existing fan mounted right there in front of you in the kitchen, there isn't a great deal you can do to quiet it down except slow it down. It is possible to leave the filters and remove the blowers from the fan and install a new external fan on the outside of the house.

If the hood has no ducting and instead simply recycles the room air through a filter, it is not a ventilating device and doesn't accomplish much by way of air quality improvement. It will collect some particulates, and it does need to be maintained regularly.

For a new installation, selecting a hood with a low sone rating is a good place to start. To help compare hoods with high airflow to those with less flow, HVI provides a "working speed" sound rating; often a high quality, 500 cfm hood is several times quieter at working speed than a 150 cfm hood. Alternatively, there are fans designed to be installed remotely from the hood and even mounted on the exterior of the house. Quiet hoods have sound insulation and anti-vibration mounts so that their motor noise does not get transferred to the building structure and amplified. As with all fans, keeping the air moving smoothly through the path to the outside will help to keep the sound level down. The effect of a convoluted ducting path is less with a range hood than with a bath fan because the volume of air is so much greater, but all the elements make up the system and by optimizing each of them the total noise produced by the system will be reduced.

SOUND CONSIDERATIONS FOR RADON/SOIL GAS EXHAUST FANS

A soil gas or radon mitigation system is designed to draw the air up from the ground surrounding the foundation of the house. Commonly these systems run inside the house with the fan mounted either in the basement or the attic, remotely from the living space. The primary consideration is that the fan mounting be isolated as much as possible from the structure of the house to minimize vibration. Airflow noise should be completely contained within the ducting.

If it is a retrofit application, however, and the fan is mounted on the outside of the house, it will be important to locate and mount the fan in such a way that operating sounds are not transmitted back into the house through open windows (see Figure 5.4). The exhaust opening from the ducting should extend all the way to the roof from the standpoint of removing the pollutant and also the fan and airflow exhaust noise above the level of the house.

FIGURE 5.4   An exterior soil gas fan. (Infiltec)

Chapter 6
INSTALLATION DETAILS

SYSTEM DESIGN OVERVIEW

Having chosen the type of system to be used, it is vital that it be installed so that it works well and is serviceable (see Figure 6.1). Remember that air is lazy and doesn't like working too hard. Duct runs should be as straight and smooth as possible. Flexible ducting is twice as resistant to airflow as smooth, metal ducting. Think about sucking your drink through a straw: the smaller and more twisted the straw the more suction it takes. Abrupt, 90-degree elbows slow the airflow and make it more turbulent. Changing those 90-degree turns to a pair of 45 degree fittings will ease the turn and thus ease the flow. Transitioning from small diameter ducting to larger ducting will also improve the flow. Use fittings that match the system and transition to a larger size, never a smaller one. Use a 4-inch duct with at least a 4-inch hood, for example. Flex duct should be pulled out to as close to its full, extended length as possible and not left bunched up in a tight corner or worse yet, bunched up in the box. If oval duct runs are needed to move air through a small wall cavity, the run should be kept as short as possible.

Keeping the air moving through the system by using larger ducting and fittings at a low velocity of 500 fpm or less (2.5 m/s) will keep the resistance lower and keep the noise level down. "The resistance of any system of ducts, grilles, filters, etc. is proportional to the square of the air velocity through it—keep velocities down by using recommended duct, grille, and filter size."[1]

To calculate the resistance of 100 feet of flexible duct, use the formula:[2]

$$Friction_{loss} = 2.74 \times (V/1000)^{1.9}/Diameter^{1.22}$$

where

$Friction_{loss}$ is in inches of water gauge;

V = the velocity of the airflow in feet per minute;

$Diameter$ = diameter of the duct in inches.

1. Vent-Axia has extremely detailed and useful ventilation information on its site: http://www.vent-axia.com/knowledge/handbook/section1.asp.
2. See the tables later in this chapter for equivalent duct lengths.

The first turn out of a ceiling mounted bath fan is the most important. Making a 90-degree turn right out of the fan will have a significant impact on the flow through the system because it is the highest pressure point in the system. Try to ease the first direction change using 45 degree turns instead of a 90 degree or move the first turn at least 4 feet away from the fan. (Fan and ducting manufacturers' instructions have good installation details. Reading them will help you install the system correctly.)

Wall cavities are commonly framed with 2 x 4 wood providing only a 3½-inch wide cavity, which is not conducive to using 4-inch ducting. "Ovalizing" or crushing 4-inch flex ducting is very resistive to airflow. Try

FIGURE 6.1   A vent in the shower. (Panasonic/Morrissey)

using hard metal, oval ducting instead. Oval, plastic ducting is available, although not common. Runs made with any of these approaches should be kept as short as possible because despite the extended area of the oval duct, it is surprisingly resistive to airflow.

Every mechanical system, no matter how beautifully engineered and manufactured, will need to be serviced at some point in its life. Homes should have a very long life—100 or 200 years or more. It should not require tearing out the 100 year parts to get to a 10 or 15-year-old component that has failed. Design and install the system so that it is serviceable. Don't bury an in-line fan in a ceiling without providing a maintenance cover. If the fan vents a first floor bathroom, think about using an externally mounted fan, remote in-line fan, or through-wall fan. Or put a grille in the bathroom and run the ducting down to the basement and out of the building from there.

Hoods and backdraft dampers also resist the flow of air, often to a surprising degree. A backdraft damper is designed to prevent outside air from coming back down the duct into the room. They are generally pushed open by the airflow and closed by gravity. Although they are quite lightweight, they still take energy away from the fan flow. They are often designed not to close completely so that a small flow of air can slide around the edge and start the damper moving when the air starts pushing in the desired direction. Without that slight opening there would be no air movement and the damper wouldn't open.

It may not be necessary to have dampers in both the fan and the external hood. Stopping the incoming flow of air at the external wall surface means that the duct won't be full of freezing air in the cold weather months, but if the damper in the hood is not effective and the damper in the fan is removed, cold drafts may occur in the bathroom. Because there are variables in every system, it is unfortunately impossible to state categorically that one damper or the other can be removed.

It is not often that the ideal ducting path can be accomplished and compromises will have to be made along the way. The goal is to keep the number of compromises as low as possible. Some fan designs are more tolerant than others to installation variations. In-line fans, for example, are specifically designed to move air through ducting and overcome resistance and since they are generally mounted remotely from the space they vent, they can be noisier (see Figure 6.2). Through-wall fans don't have any extra resistance to deal with and so their designers can engineer the performance characteristics in the laboratory.

Air moving through a duct is larger, slower, lazier, and has less energy than electrons moving through a wire. If there is a design decision to be made

FIGURE 6.2 A remote mounted fan. (Fantech)

about whether the path for the ducting or the path for the electrical wire should be shorter or straighter, the ducting should always win out. A few extra feet of wire may cost a few extra pennies, but the electrons generally don't care (on the scale of a house). Air, on the other hand, does care. Keep the runs as short, straight and smooth as possible. And never, ever vent a bath fan into an attic. It will deposit excessive moisture in the attic and cause numerous problems. Ventilation systems must be vented all the way to the outside—not just close to it (see Figure 6.3).

DUCTING

Types of ducting

Ducting is available in a variety of styles for a variety of purposes. Never use that white, uninsulated flexible ducting that is sold as a dryer vent duct for a ventilation system of any kind or purpose. It's like using duck tape on ducts! It should never happen.

TABLE 6.1 Duct areas.

Smooth Round Duct Diameter	Free Area (Square Inches)
2″	3.14
3″	4.71
4″	6.28
5″	7.85
6″	9.42
7″	11
8″	12.57

FIGURE 6.3 Never vent a bath fan into an attic. (Panasonic/Morrissey)

Round, galvanized, 24-gauge sheet metal ducting comes in standard, 4-foot lengths "un-locked" so that it can be cut to length. One end is crimped so it will fit into the next section of ducting. The crimped ends should be oriented away from the source of pressure (the fan), otherwise it is like installing a roof from the bottom up! There are all sorts of fittings—elbows, "Y's" (or Wyes), "increasers," "de-creasers," etc. It's smooth and ventilation air moves easily through it with little resistance (see Figure 6.4). There are a lot of joints, however, each of which needs to be carefully sealed to prevent losses. It's also un-insulated so that warm moist air moving through it will condense in it when the duct passes through cold, ambient conditions such as a cold winter attic. Insulating ducting blankets are available and can be wrapped around the pipe. The blanket has a vapor barrier, which should be kept on the outside and sealed tightly along the seam and at the joints to make it continuous. Ducting in an attic can be covered by the attic insulation or foamed over. Just bear in mind any service issues that may need to be dealt with in the future. Don't make mechanical devices inaccessible.

White, Schedule 40, PVC sewer pipe comes in much longer lengths, which reduces the number of joints that must be made and sealed, and, as can be seen from Table 6.2, it is more than 100 times smoother than flexible ducting. Limiting the joints makes it even smoother. It too will need to be wrapped or covered in insulation, however, when it passes through a space outside the heated envelope of the house.

Oval, galvanized pipe is commonly used for heating or air conditioning system risers or stacks, passing up through wall cavities. Although it is considerably less resistive to airflow than ovalizing round, flexible ducting, it is

Pressure is listed in "inches of water gauge" (w.g.) on the inch/pounds scale and in Pascals (Pa) on the metric scale. The two are getting blended in the building science business. Most residential ventilation systems experience pressures of much less than 1 inch w.g. or 249 Pa. Most bath fans are rated at 0.1 inch w.g. (24.9 Pa) and some ventilation codes require a performance rating at 0.25 inch w.g. (62.2 Pa). As fabric duct gets larger in diameter, its ability to handle higher pressures is reduced and requires tougher material to achieve the same rating. Lower pressure handling ducting is rated by UL as "connector," applicable only for short runs. Higher-pressure capable materials get defined as "duct." Ventilation systems can use the "connector" grade because of the low pressures involved.

FIGURE 6.4   Flow in round sheet metal duct. (Panasonic/Morrissey)

TABLE 6.2   Duct roughness.

Duct Material	Roughness Category	Absolute Roughness ε, foot
PVC plastic pipe (Swim, 1982)	Smooth	0.0001
Galvanized steel, longitudinal seams, 4 ft joints (Griggs, et al. 1987)	Medium smooth	0.0003
Semi-flexible duct, metallic, when fully extended	Rough	0.007
Flexible duct, all types of fabric and wire	Rough	0.01

Table from ASHRAE Handbook of Fundamentals 2005, p. 35.7.

still much more resistive than round ducting of a similar cross sectional area. For the equivalent of a 4-inch diameter galvanized duct you would need to use a 3 inch by 6 inch oval to achieve the same area.

Metal semi-flexible ducting is smoother than the fabric type because the joints have a lower profile. It needs to be extended out to its full length for optimum performance, and it also needs to be insulated. This ducting does have a tendency to get damaged fairly easily as well and should not be located in a vulnerable spot. When it is dented, it will not recover on its own.

Non-insulated flexible duct has the same flexibility as the insulated variety, but it does not include the insulation blanket. It comes in a variety of styles and qualities. The white vinyl stuff that is supplied for dryer vents is at the low end of the quality scale. Smaller diameters (less than 4 inches) are often qualified as "hose" rather than ducting. Underwriter Laboratories (UL) has test procedures for flexible ducting that qualify it as either "air duct" or "air connector," suitable only for shorter runs. When flexible ducting is used in commercial applications where there is a significant amount of pressure in the duct, a large amount of air is forced in a small diameter space. Ventilation air velocities and pressures (particularly in residential applications) are not high pressure, and the air is moving at low speeds. Therefore it is not necessary to use "high pressure" ducting. Increasing the volume and increasing the velocity of air movement will increase the noise of the system.

Flex duct manufacturers make ducting of a trilaminate of aluminum foil, fiberglass, and aluminized polyester or PVC coated fiberglass cloth or chlorinated polyethylene (CPE) materials. Air duct rated flex duct can handle as much as 10 inch w.g. through 12 inch diameter ducting. They are rated for velocities in the range of 5500 feet per minute (fpm). (Remember for ventilation the velocities should be below 500 fpm to keep the sound level down.) Just to confuse things a bit more, the UL ratings between "air duct" rating and "air connector" rating depend on the number of tests that the manufacturer has put the product through. Air duct must pass all the tests of the UL 181 Standard. Air connector must pass only a limited number of tests and is limited to "installation in lengths not over 14 feet." It may be that the manufacturer doesn't think the air connector product will pass the full range of tests or it may be that the manufacturer simply doesn't want to invest in all that testing. For residential ventilation systems, however, it doesn't matter whether it is air duct or air connector, because unless something very strange is being installed (see your qualified engineer), the velocities and the pressures will not get high enough to be a concern.

Insulated flex duct is flexible duct wrapped up in an insulating blanket. Typical insulation levels are R4.2, R6, and R8. The insulation is commonly fiberglass held in place with a vapor barrier of metalized polyester (if it is the silver version), a black, polyethylene covering, or a scrim reinforced polyester jacket. To make the vapor barrier work as designed, both the ducting and the vapor barrier need to be sealed at the ends. If only the ducting is sealed and the vapor barrier is left open, the cooler or hotter air will successfully penetrate under the vapor barrier, and it will lose its effectiveness. Fiberglass insulation is not good at blocking airflow that sneaks into the open ends of the of vapor barrier. If moisture gets under the vapor barrier and condenses on the ducting, it will eventually saturate the fiberglass, eliminating the insulating properties, and pooling in the vapor barrier with the ensuing potential disasters.

The Air Diffusion Council has some excellent installation details, much of which is included by manufacturers in their installation instructions.[3] Their installation details include how to attach the ducting to fittings, making splices, and what tapes, duct mastic, and connectors to use. The details are primarily for HVAC or higher-pressure ducting but should be applied to good ventilation system installations as well.

Flexible ducting needs to be supported along its length (see Figure 6.5). Sags and dips increase flow resistance (each one is a series of turns), and each dip is a potential moisture reservoir where water can collect and increase the dip.

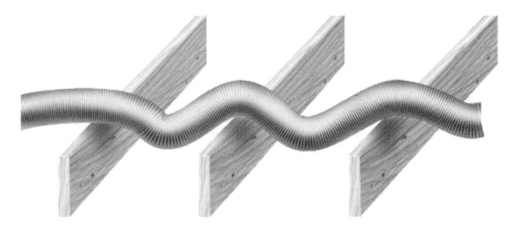

FIGURE 6.5   Flex ducting must be supported. (Panasonic/Morrissey)

3. Air Diffusion Council, Schaumburg, IL, http://www.flexibleduct.org/ADC_Inst.asp.

Duct Joints, Mastics, and Duct Tapes

Elbows, Y's (Wyes), and T's are all places where the air is forced to change direction. Making direction changes with hard pipe fittings is less resistive than making similar direction changes with flexible ducting. Keeping the air moving in a straight line is the best way to keep it moving with the least resistance. A straight line from the outlet of the fan to the outside is not a reality, however, but achieving the straightest, smoothest path—minimizing dips, sharp bends, and crushed ducts—will provide the best performance.

Joints that are factory made are considerably tighter than those that are made in the field. Ideally the seam in a 4 foot length of galvanized pipe should be sealed with mastic, but if it comes down to what to seal and what not to seal, in-field-made rather than factory-made joints should be addressed first, working out from the highest pressure points near the fan, HRV/ERV, or air handler.

In an HVAC system, there are numerous bends and turns as the air moves from the central air handler, commonly through a "trunk duct" with "take-offs" or paths out to the individual rooms. Those take-offs leave the main trunk at an angle, commonly 45 degrees, as the branches of a tree. Turning vanes may be used to guide the fast moving air around corners, and help the air move smoothly and quietly from its source to its destination.

The installation of a stand-alone HRV or ERV system requires many of the same design considerations. Air is moving from the outside to the HRV or ERV, from there to the rooms, from the rooms back to the HRV or ERV, and from there back to the outside. To keep the system operating smoothly, care needs to be taken in the design and installation of the ducting. Slamming air into a right angle turn will create internal turmoil in the flow and slow it down far more than if the air is coaxed into changing direction through a series of 45 degree turns.

FITTINGS: Fittings are metal duct elbows that come segmented so they can be adjusted from a 90-degree turn to a straight line (see Figure 6.6). One end is crimped to fit into the next section. The crimped end of any duct section or fitting should be at the low-pressure end, away from the fan to lessen the leakage. Many fittings are more commonly used for HAC (heating and air conditioning) applications where the air is moving out from the air handler to the rooms so the crimping may be on the wrong end for ventilation, which moves the air from the rooms to the outside.

"Increasers" increase the diameter of the ducting, which will reduce the pressure and improve the airflow (see Figure 6.7). They can increase the diameter by one or 2 inches. When they are tapered, the air flows easily from one size to the other. Un-tapered increasers save on size, but the transition is

FIGURE 6.6   A 90-degree elbow. (Morrisey)

FIGURE 6.7   An increaser and a reducer. (Morrisey)

an abrupt "T" so they will increase turbulence at the fitting. Although these un-tapered "increasers" can also be used to decrease the duct size, the friction losses will be very high.

"Y" (or Wye) fittings can bring two duct runs together, blending them into a single and generally larger duct. Y fittings can look just like the letter or they can be "Tee-Wyes" that look like a straight duct with a branch entering at the side. There are also "Tri-Wye" fittings that have three connections in and one connection out. If it is desirable for the airflow from the inlets to be equal, then the duct diameters going into the "Y" should be equal and the duct diameter going out should be greater—4 inch x 4 inch x 6 inch for example. Reducing the size of one of the inlets will reduce the flow through that inlet. This might be desirable in tapping off a master bath and a powder room or the middle of the bathroom and a toilet space.

"Collector boxes" can join the ducts from a number of rooms together into a box with a single outlet that is then vented to the outside. This allows for a number of rooms to be vented to the outside through a single fan and a single opening in the building envelope. All the rooms will be vented simultaneously unless a controllable, dampering arrangement is added to the system.

MASTICS: Mastics are flexible sealants that vary in consistency from yogurt to mashed potatoes. The sealant remains flexible over an extended period of time, shrinking and expanding with the changes in duct temperature. It is sold in tubs, buckets, or cartridges like caulk. It is either solvent or water based. In terms of working in confined spaces, the water-based caulks are preferable for the health of the installer. Mastic can be applied with a putty knife, stiff brush, or even fingers.

According to Oikos[4] there are a number of elements to look for in good mastic:

- *High solids content*—Solids content listed on the product literature should be at least 50 percent. Some have as much as 70 percent. Generally, higher solids means less shrinkage as the material cures.

- *Excellent adhesion*—The mastic must stick to metal, wood, drywall, plastic, concrete, and just about any other material that might be found in a house. Since these surfaces are seldom clean, look for mastics that hold well to dirty or oily surfaces.

- *Excellent cohesion*—The mastic must also stick to itself so it doesn't crack as the surface moves or "blow out" under the air pressure inside the duct.

- *Water resistant*—Condensation may collect on ducts during cooling, so the mastic must hold even when exposed to water.

4. Iris Communications, http://www.oikos.com/library/ducts/index.html.

- *Non-toxic*—Check the material safety data sheet for warnings. If you choose to spread the material with your hands, you may need to wear gloves. Water-based mastics are generally less irritating than petroleum-based ones.

- *Surface burning characteristics*—Duct mastics should be tested for flame spread and smoke developed (UL723 and ASTME-84). National Fire Protection Association Standard 90A requires duct mastics to have a flame spread rating no higher than 25 and a maximum smoke developed rating of 50.

- *Viscosity*—Some installers like thicker mastic that they can apply with a trowel or gloved hand. Others like it a bit thinner, so it brushes on easily. Viscosity is measured in units called centipoise (cps). Higher numbers mean thicker mastic. Viscosity around 100,000 cps indicates a consistency similar to mashed potatoes, while 60,000–75,000 cps suggests something closer to yogurt.

- *Storage*—Many manufacturers emphasize that their products have a limited shelf life, usually about one year. Most mastics should not be allowed to freeze during storage or shipping.

- *Color*—If ducts will be visible, you may want to select a pleasing color.

- *UL Listing*—Mastics are tested by Underwriter Laboratories to Standard 181-B and may be marked as 181B-M on the container.

Because there is no standardization of these products, not all of this information will be comparatively available.

TAPES: "Duck" tape, the cloth kind with the adhesive backing, is good for just about everything except for ducting. It was originally made of a duck fabric and used as a weatherproof joiner in the military. (The "duck" name came both from the fabric and from its waterproof qualities.) It wasn't until after World War II that it changed to grey or silver and was attached to duct work. Now it comes in an enormous variety of colors, has dozens of books and myths surrounding it, but it is only mentioned here in passing because it shouldn't be used for sealing ducts. It dries out quickly and falls off.

There are, however, many other tapes that are effective and long-lived for sealing the joints between ducts, connectors, equipment, and each other. It should be noted that tapes and mastics are <u>not</u> designed to mechanically attach or support duct connections. Their purpose is to stop leakage.[5]

Underwriter Laboratories (UL) has several tests for tapes: UL-181A for rigid ducting, UL-181B for flexible air ducts and connectors, and UL-181B-FX for adhesive tapes on flexible ducting. UL-723 is for surface burning of building materials. Manufacturers have all sorts of "grade" designations such as *economy, utility, general purpose, contractor, industrial, professional,* and

5. LBNL-41434. "Can Duct-tape Take the Heat?" Max Sherman & Iain Walker.

premium, but even with these tests and labeling it is difficult to know what tapes will function effectively over an extended period of time. If the ducting is to be "buried" in a wall or building component that will be inaccessible without extensive destruction, mastic is a much better choice for sealing the joint. Ventilation ducting is not subjected to the same variations in heating and cooling that HVAC ducting is exposed to. Nevertheless, it is important that the air not leak into or out of ventilation ducting. Ventilation air is the air that the occupants are relying on to breathe.

The color of the duct tape is not an important component of its ability to do its job. It may be more attractive to join pieces of silver ducting together with silver tape, but using clear tape will allow the installer or inspector to see the joint and determine if the gaps have been completely covered. Clear, UL-181B tape is available.

Installation details are as important, or perhaps more important, than the tape itself. Understandably in an attic, basement, or crawl space it is virtually impossible to make the joint clean and free of dust, but the closer the installer can get to that the better. From a duct sealing standpoint, the UL rated tapes don't have an advantage, but local codes may require their use.

Ducting Routes and Layout

The airflow through the ventilation system depends on the layout and installation of the ducting and fittings. When an air stream changes direction, a dynamic loss occurs. Unlike friction losses in straight ducting, fitting losses are due to turbulence rather than skin friction. It is not always easy to calculate or measure the static pressure of the system. The resistance to the airflow can be estimated by using the *actual length*, the *equivalent length*, and the *effective length* as stand-ins for *static pressure*.

* *Actual length*: The actual, physical length of the duct run (double the *actual length* to estimate the resistance of flex duct run);
* *Equivalent length*: The equivalent of resistance to airflow for a specific fitting;
* *Effective length*: The total of the *actual length* and the *equivalent length* of all the fittings in a run.

The airflow of a fan or ventilation product is certified at a specific static pressure (0.1 inch w.g. for ducted fans like bathroom exhaust fans or whole house comfort ventilators, 0.03 inch w.g. for direct discharge products without ductwork, 0.2 inch w.g. for in-line fans, and 0.1 inch w.g. for kitchen range hoods) although in the process of being tested, the performance is measured at a variety of pressures from zero to the maximum resistance the fan

Installing a ventilation product and attaching ducting without understanding how much resistance there is in the ducting and fittings guarantees that the airflow will be severely diminished. The common industry rule-of-thumb is that the flow will be about half of a fan's certified flow-rate. So you could simply buy a fan that moved twice the air required and ignore the installation issues. What is surprising about going through these steps is how much resistance is generated by the fittings and how quickly most ventilation products will reach their performance limit. The best way to know how the installed product is working is to test it or measure it.

can handle when the flow goes to zero cfm, generating a "flow curve." To accomplish that flow once it is installed in the house, the ventilation product has to be installed so that the static pressure point or flow resistance at which it was rated is not exceeded. If it is exceeded, the installed flow will be less. Because most installations are not carefully done and do not consider the laziness of the air motion, it is common to find most ventilation installations are operating at about one half or less the rated flow.

An ERV or HRV layout is similar to the layout for an HVAC system. Both the fresh air supply and the stale air return have to go out to the rooms and back to the heat exchanger. The resistance to the flow through each of those paths should be equal so that the same amount of air moves in both directions. For an HVAC system, balancing the ducting design means that when the conditioned air reaches the grille in the room it will be moving at a velocity that will satisfy the needs of the room for conditioning the air, and it will be moving quickly enough through the grille to be "thrown" out into the room in a way that will stir up the air throughout the room. The velocity of the airflow through the grilles of an HRV/ERV installation or a supply-only ventilation system should be equally carefully considered so that the occupants are not subjected to drafts.

Calculating the Effect on Performance

For an exhaust fan the ducting layout must go directly to the outside through the smoothest, straightest possible path (see Figure 6.8). This is true for a bath fan, kitchen fan, or range hood. Minimizing the resistance will maximize the performance.

1. Select the desired flow rate or cfm for the fan (for example 60 cfm).
2. Determine the location of the fan.

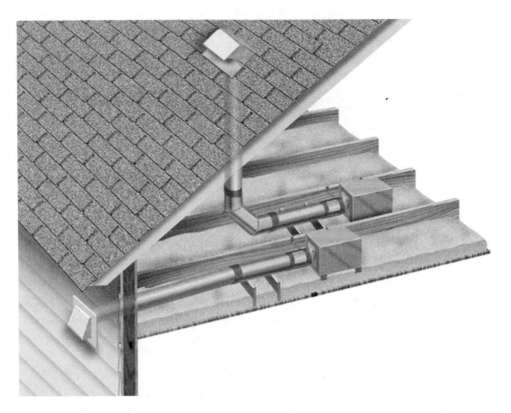

FIGURE   6.8   Ventilation systems must vent all the way to the outside. (Panasonic/ Morrissey)

3. Orient the fan so that the fan's outlet is pointing toward the outside of the building whether that means exiting through a wall or through the roof.

4. Choose the starting type and size of ducting to be used from Table 6.3. (For example, 4 inch insulated flex duct for a duct run in an attic to the roof for a 60 cfm fan will be able to move that much air up to a total *effective length* [the total of the *actual length* of the duct and the *equivalent length* of the fittings] of 20 feet. If rigid ducting were used, the *effective length* could be increased to 40 feet.)

5. Determine the path or "outside duct run" between the fan outlet and the vent or hood, determining the number of 90-degree and 45-degree elbows. (Use Table 6.4 for the *equivalent length* of the fittings.) (For example, the 60 cfm fan will be installed in the ceiling of the bathroom. The flexible ducting will run for 2 feet before being turned up 90 degrees headed straight up 4 feet to the roof where it will attach to a low profile roof cap

TABLE 6.3 Maximum effective flexible duct length at 0.1 inch w.g.

	3 inch diameter	4 inch diameter	5 inch diameter	6 inch diameter	8 inch diameter	10 inch diameter	12 inch diameter
30 cfm	17 feet	75 feet	225 feet	NL	NL	NL	NL
45 cfm	8 feet	35 feet	100 feet	250 feet	NL	NL	NL
60 cfm	DU	20 feet	60 feet	150 feet	NL	NL	NL
75 cfm	DU	13 feet	40 feet	100 feet	400 feet	NL	NL
90 cfm	DU	9 feet	28 feet	65 feet	300 feet	NL	NL
105 cfm	DU	DU	20 feet	50 feet	225 feet	NL	NL
120 cfm	DU	DU	15 feet	40 feet	175 feet	NL	NL
135 cfm	DU	DU	12 feet	30 feet	130 feet	400 feet	NL
150 cfm	DU	DU	10 feet	25 feet	110 feet	350 feet	NL
165 cfm	DU	DU	DU	20 feet	95 feet	275 feet	NL

DU = Don't use, NL = No limit (greater than 400 feet).
Effective lengths for rigid ducting can be double these lengths.

with a backdraft damper and bird screen. If the 90-degree turn was made with two, 45-degree metal, adjustable elbows, each elbow would have the *equivalent length* of 5 feet (a total of 10 feet). If the 90-degree turn is made using the insulated flex without the metal elbows, it would have the *equivalent length* of 20 feet (doubling the 90-degree metal fitting length from Table 6.4).

6. Select an exterior vent or hood and note the *equivalent length*. (Use Table 6.5 for the *equivalent length* of ducting for the vent.) (For example, a low profile roof cap with a backdraft damper and bird screen will be used for this installation because that's what the owner wants, with an *equivalent length* of 60 feet.)

   Total *equivalent length* of this installation:
   Flex duct 90-degree elbow   20 feet
   Vent   60 feet
   Total *equivalent length*   80 feet

7. Calculate the *actual length* of ductwork to be used, in this example 2 feet plus 4 feet or 6 feet of rigid duct, doubled to 12 feet for flexible ducting.

8. Add the *actual length* to the total *equivalent length* for the elbows and the vent to determine the overall *effective length* and use Table 1 to determine the duct size to be used for the required airflow rate.

   Total *effective length* in this example:
   Total *equivalent length*   80 feet
   Total *actual length*   12 feet
   Total *effective length*   92 feet

TABLE 6.4   Fittings—equivalent lengths (rigid ducting)[6].

Image	Description	Equivalent Length (ft) for Rigid Ducting[7]
	45-degree adjustable elbow 2-piece	5
	90-degree adjustable elbow 4-piece	10
	Wye, equal sizes	10
	Tee, take-off	50
	Tapered increaser/reducer	4
	Hard increaser/reducer	8

6. Adapted from HRAI "Residential Mechanical Ventilation."
7. These are equivalent lengths of rigid ducting. Flexible ducting equivalent is twice this amount.

TABLE 6.5   Vents, hoods, and wall caps—equivalent lengths.

Image	Description	Equivalent Length (ft)
	Triangular wall cap for round duct with backddraft damper and bird screen	60
	Triangular wall cap for round duct with bird screen, without backddraft damper	35
	Rounded wall cap for round duct with backdraft damper and bird screen	40
	Louvered wall cap	50
	Low profile soffit vent with backdraft damper and bird screen	60
	Roof cap, low-profile for round duct with backdraft damper and bird screen	60
	Roof cap, "goose-neck," for round duct with backdraft damper and bird screen	35

To move 60 cfm through 92 feet *effective length* of ducting at 0.1 inches of static pressure would require using 6-inch diameter flexible duct (150 feet in Table 1). Four inch ducting would be too small, allowing only 20 feet of *effective length*.

9. If the resistance exceeds the capability of the fan, use the next larger size duct, reduce the number of elbows, select a better exterior vent or a fan with a higher performance. The fan's rated performance is measured at 0.1 inch w.g. Some fan performance curves are also certified at 0.25 inch w.g. Will the fan be able to deliver the required flow for the installation?

From this exercise it is easy to see why most installations do not deliver the rated amount of flow. Flexible ducting less than 6 inches in diameter is too resistive for almost any length of duct run with a fan that is rated at 0.1 inch.

HRV/ERV duct design: There are two basic choices for designing the distribution system for an HRV/ERV. The first is a stand-alone system with ducting running to the living spaces and exhausting from the "polluting" spaces like the bathrooms and the kitchen. Designing the ducting for this is similar to designing the ducting for an HVAC system. It should be carefully thought out and the installation must be balanced for the system to work correctly. The description here is brief and only touches on the major points.[8]

The "equal friction method" is a relatively simple process for designing a balanced ducting system for an HRV/ERV. It requires calculating the amount of airflow that needs to be delivered or removed from each room, totaling the individual room requirements to determine overall flow, and calculating the total system pressure drop using the duct sizes and flow resistance. It is important that this process be done correctly. If the system is unbalanced, it will not work correctly and can actually do more harm than good in some cases. The complete stand-alone ducting design process is beyond the scope of this book.

The second approach is to integrate the HRV/ERV with the existing ducting of the HVAC system. This is a common but much less satisfactory approach. The system will still need to be balanced. A compromise to this approach is to draw the exhaust air directly from the polluting points (like bathrooms) and just use the HVAC system to deliver the fresh air.

---

8. HRAI conducts two-day courses in ventilation system design and has detailed design and installation information through its SkillTech Academy: http://www.hrai.ca/site/skilltech/home_skilltech.html.

TERMINATION FITTINGS

Exterior Hoods and Vents

The termination fitting or exterior vent or hood is one of the most overlooked and taken-for-granted components of the ventilation system. The air from the ventilation system has to pass through the building envelope and leave (or enter) the building (see Figure 6.9). That penetration must not allow water or creatures to enter, and it should only allow air to enter or leave when the ventilation system is running. And it must accomplish all of this with as little resistance to the desired air movement as possible.

To keep the water out, roof and wall vents make a sharp, 90-degree turn at the wall surface or roof. The sharper that turn, the more resistance that the

FIGURE 6.9   Never locate an outlet near an inlet. (Panasonic/Morrissey)

vent creates. Vents that have a rounded turn ease the air into the new direction and are less resistant to the flow (see Figure 6.10).

Most exterior vents include an airflow operated backdraft damper or flap that blows open and closes with gravity. (On some vents, the damper is removable so that the vent can be used as an intake.) Despite the fact that the damper material is light, it takes energy to push it open. That energy is subtracted from the airflow. Some flaps can reduce the flow by as much as 50 percent! If there isn't much airflow by the time it reaches the vent, there won't be much energy to push the flap open at all.

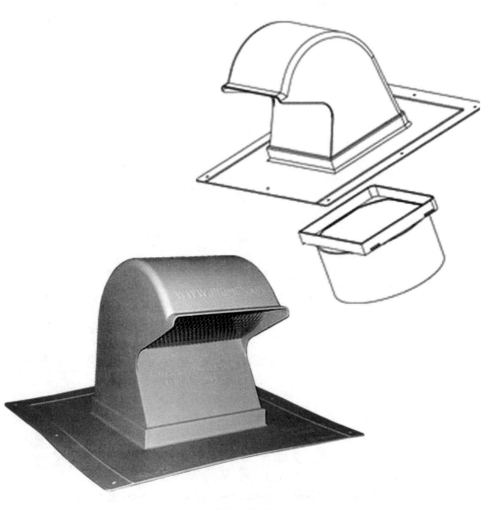

FIGURE 6.10 A roof mounted exhaust hood. (Primex)

The outside surface of the home is the ideal place to block the incoming flow of outside air, keeping the interior ducting at room temperature. Most ceiling mounted bath fans have built in backdraft dampers. If there is a damper in the fan, the second damper in the exterior vent can be removed and the damper in the fan will block flows of outside air into the room, but as mentioned previously, since all installations are different a decision needs to be made at the installation site. Airflow performance will be improved.

For a remote mounted or in-line fan that does not include a backdraft damper, the damper in the exterior vent serves the purpose reasonably well. If greater protection is required, a separate backdraft damper in the line can be used to replace the exterior flap.

For an HRV or ERV there can be no airflow-operated flap in the intake side to prevent air from coming in. A motorized damper can be used if it is desirable to seal the system off when it is not running.

Low airflows or no airflows through the vent (particularly wall vents) will provide an ideal spot for nesting birds (see Figure 6.11). They quickly learn to flip open the damper and climb inside where it is warm and dry and make a nest. And when the baby birds grow up and leave, the mites that live in their feathers may seek out other homes farther into the house. It is a good idea to use an exterior vent or termination fitting that has a creature screen. These are too broad to cause much resistance to flow and serve their protective purpose well. (Insect screens, on the other hand, are extremely resistant to airflow and should only be used with powerful fans and in conditions where insects are a major problem. Running the fan frequently or constantly solves the problem better.)

Grilles, Registers, and Diffusers

Ceiling mounted bath fans include a grille. The certification process requires that they be tested with the grille installed so the grille is engineered to match the product and is figured into the airflow specification.

In-line or remote mounted exhaust fans, however, generally do not come with a grille, and a variety of options are available. Most fan manufacturers will offer grilles to match their fans. In calculating the *equivalent length* of duct for the resistance of the grille, use 10 feet for a relatively open, round grille and 15 feet for a stamped metal louvered grille.[9]

If the rate of airflow is low relative to the size of the opening, the grille will have less resistance and make less noise. At higher velocities, grilles begin to create noise in the room as the air rushes through the opening. An open, low resistance, well-designed grille will make the system operate more quietly.

---

9. See Chapter 9 for more information on grille resistance.

FIGURE 6.11 A bird's nest located in a louvered hood. (PHR)

If an HRV, ERV or supply-only ventilation system is integrated with the home's HVAC system, the supply grilles will be installed to meet that system's requirements. Supply diffusers for dedicated ducting HRVs, ERVs, and supply-only systems must be designed so that the air leaving them is moving at a high enough velocity to "throw" the air out into the room and satisfactorily mix with the room air. Discharge velocities between 500 and 800 fpm are a reasonable compromise between air mixing, noise, and pressure drop. In a cold climate, the air leaving these diffusers will be at a cooler temperature than the room air so they need to be installed high enough so that they are not creating drafts that the occupants feel are uncomfortable. If the diffuser is located on the floor, the incoming air temperature should not be less than 60°F. If the diffuser is located high on the wall or on the ceiling, the incoming air temperature should not be less than 54°F. Many grille manufacturers have performance data on their products.

DAMPERS

Balancing Dampers

HRV and ERV systems need to be balanced—equal amounts of air supplied and exhausted to and from the house. A carefully designed and installed ducting system will do most of that work, but balancing dampers should be included to finish the task. A typical balancing damper is an adjustable vane in the air stream that can be manually adjusted to block some portion of the flow by rotating the damper to be parallel or perpendicular to the flow with a small handle located on the outside of the ducting. An alternative balancing damper is constructed like an iris that can be manually opened or closed to vary the amount of air passing through it.

A manometer is used to measure the flow in the two air streams while the adjustments are made then the balancing dampers can be locked in position to maintain that consistent airflow.

Backdraft Dampers

Most of the time air should flow through a duct system in only one direction. Exhaust ducts should only exhaust. Supply ducts should only supply. Backdraft dampers like the flap on an exterior hood are designed to blow open with the flow and fall back to the closed position when the flow stops. They are generally flat, like pieces of metal or plastic that don't take a great deal of force to push open but have enough weight to close. Some fan manufacturers counterbalance their dampers to make them more rigid and to close more positively.

Using the HVAC ducting for fresh air distribution from an HRV or ERV is common practice, but it has significant drawbacks. First, for the system to effectively distribute the ventilation air, the blower in the air-handler must run at the same time as the blower in the HRV/ERV. This is a significant operating cost and energy penalty. The fans in the HRV/ERV are not strong enough to effectively distribute the air through the much larger air-handler ducts. Second, in the "swing" seasons when the air-handler is not running regularly and ventilation is most needed, the air-handler is less likely to be running. Third, in an air-conditioning dominated climate, the humid air drawn in through the HRV/ERV may cause condensation in the HVAC equipment. These surfaces will be around 60°F and outdoor dew points are likely to be commonly well above that temperature. Fourth, it is almost impossible to balance the system. If the HRV/ERV is trying to draw air out of the return side of the system the negative pressure at that point may be high enough to stall the HRV/ERV fans.

The duct attached to an exhaust fan is purposefully designed to provide a low resistance "hole" to the outside of the building. If it has been properly designed and installed to optimize the performance of the fan, it is akin to cutting a 4, 5, or 6-inch hole in the side of the house. The damper is the only thing that blocks the flow of cold air in winter or hot, humid air in summer from coming back down into the house. Although its *conductive* insulation value is generally so low it is not worth considering, its *convective* resistance (its ability to block air movement) should be high.

Most backdraft dampers installed in fans work quite well at blocking the back flow of air when they are new. When a fan is labeled "For ceiling installation" it primarily means that the damper has been balanced to be installed in the ceiling. When it is closed, the damper rests on a surface in the "nozzle" of the fan. There is often a "bumper" or rest on that surface that keeps the damper very slightly open at all times so that airflow can be established when the fan starts so the damper will open, but even with that designed opening, backflow leakage is quite tight.

When the fans are performance certified to the HVI 916 Standard, their flow is measured with their integral backdraft dampers in place. Additional backflow prevention in the duct or in the hood itself will add considerable resistance to the flow.

A motorized damper can be much tighter when it is closed and does little to resist the desired flow because a motor pushes the damper open and holds it there rather than using the moving air for that purpose (see Figure 6.12). Motorized dampers need to have a control interlock with the fan so that both devices turn on and off at the same time. Motorized dampers may work against an internal spring so they only use power when they are opening and use the spring to push them back into the closed position. This arrangement means that shutting off the power to the damper will make it close. Without the spring, the damper will need a constant source of power—power to open and power to close. It will be a small amount of power, but it complicates wiring.

Although more common in commercial applications, some dampers are operated by pneumatic pressure, a small pump providing air pressure to move the damper blade.

VENTILATION CONTROLS

Without a means to turn it on or off, a fan is just a pile of metal and plastic parts. It can be installed so that it is "hard-wired" or permanently connected

FIGURE 6.12 An open motorized damper. (PHR)

to a power source so that it is never shut off, running 24 hours a day, seven days a week. This eliminates the need for a control and for any backdraft protection. If the fan is quiet enough and energy efficient enough, this can be the optimum strategy.

Generally homeowners like to have some control over how their house is operating. It has been a common practice to wire the exhaust fan in the bathroom to the light switch, an "occupancy" sensor of sorts—when the bathroom is occupied the fan and the light are on (see Figure 6.13). Moisture stays in the bathroom, however, for a much longer time than that. It runs down the walls after a shower. It stays in the towels, and it hangs in the air. In most bathrooms, the relative humidity can remain above 90 percent for two or more hours after a shower.

A delayed off timer is a reasonable approach to the spot ventilation needs of a bathroom application, particularly if the delay can be extended to an hour or two and the fan is quiet and energy efficient. Some light switches are available that allow the light to be turned off while the fan continues to run (see Figure 6.14).

FIGURE 6.13   A labeled ventilation system on/off control. (Panasonic/Morrissey)

Another approach to controlling the ventilation with time is to install a cycle timer, a control that can be programmed either mechanically with "pins" or electronically to turn the fan on and off periodically throughout the day. These can be set up to cycle the fan to run for 10, 20, 30, or more, minutes per hour. Some of them have multi-speed capability so that they run continuously at a low speed and are boosted to a higher flow rate during times of occupancy. This strategy allows the fan to serve multiple purposes—as the continuous, low volume primary ventilation system and the higher volume the occupant may feel is necessary to clear the air (see Figure 6.15).

It would seem that running the fan on relative humidity with a dehumidistat would be an acceptable approach. Relative humidity is not as familiar as temperature, and people have a tendency to equate the two, thinking seventy percent relative humidity is comparable to 70°F. The ideal living conditions are between 30 and 55 percent relative humidity, varying by the time of year and the climate, factors that also make it difficult for a dehumidistatic control to work effectively. Small changes in the humidity settings will cause significant changes in fan run time, from minutes to hours. Without microprocessor technology the control will operate the fan when the ambient humidity is high, in some cases causing the fan to run all the time in summer. Controls do exist that

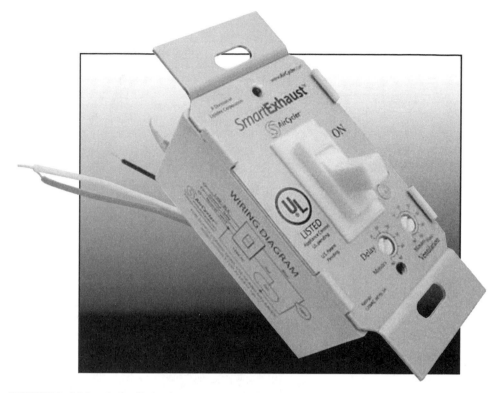

FIGURE 6.14   A fan/light timer control. (AirCycler)

respond to rapid changes in humidity rather than a natural, steady rise. A shower is a humidity event that will spike the sensor and activate the fan.

There are also sophisticated controls that operate on what is characterized as "air quality." These are "mixed gas" sensor controls to sense a variety of gases, particularly those found in cigarette smoke such as hydrogen, ethanol, iso-butane, carbon monoxide, and methane. The difficulty with these controls is that they operate when the level of invisible gases is high even when the occupant can't smell or sense the contaminant. That can be a good thing, but it is difficult to know why the ventilation system is running or if it should be.

"Demand" controlled ventilation generally uses a CO_2 sensor. Carbon dioxide is a good indicator of human occupancy of a space. Higher levels of CO_2 are a reasonable indicator of higher occupancy and the need for higher ventilation rates. CO_2 levels under 800 parts per million (ppm) are common in outdoor air conditions. These levels inside may mean the house is being over-ventilated. Between 800 ppm and 1200 ppm are normal. If the air con-

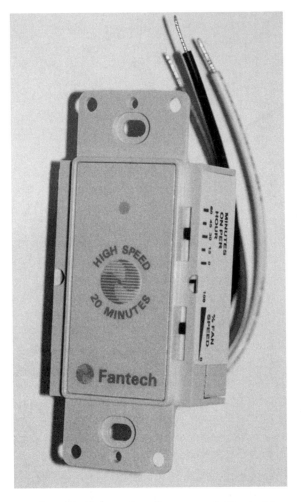

FIGURE 6.15   An adjustable flow and cycling control. (Fantech)

tains over 5000 ppm there are probably other nasty gases in the air and the space should be cleared of people until the air is cleared.

REPLACING AN EXISTING BATHROOM FAN

The first step in replacing a bathroom fan is to determine the ductwork path to the outside of the house. Is there ducting? (Installers have been known to install a fan in the ceiling but not connect it to anything.) Is ducting that runs through an attic insulated? Is the path as straight and smooth as possible?

How about the exterior vent? Does that need to be replaced as well? (Check the vent for old bird nests.)

Determine how much airflow is required. If the fan is to be used as the primary ventilation system, running continuously at a low flow, use the table from ASHRAE Standard 62.2 found in Chapter 3 of this book.

Kits are available that can be added to an existing ceiling mounted bath fan that can use the existing grille and housing and add an in-line fan to up-grade the old product to provide better performance. These products generally require removing the existing fan blade and motor and adding the new, in-line fan remotely. The advantage to this approach is that there is no impact to the bathroom ceiling. The wiring will need to be moved so that the switch in the bath will be connected to the new fan. In-line fan designs are generally powerful enough to be able to handle the resistance of the existing housing.

Alternatively removing the housing in the bathroom ceiling and adding a new grille and a remote fan will work equally well. (It will just require repairing the ceiling.) There are also exterior vents that include a fan, taking all the fan noise out of the building. The ducting from the bathroom to the exterior of the building still needs to be checked for insulation and a direct and smooth path to the exterior.

There are a wide variety of choices for replacing an existing ceiling mounted fan with another ceiling mounted fan. There are fans alone, fans with lights, fans with heating elements, and fans with lights and heating elements. Many of them have the Environmental Protection Agency's Energy Star® label which means that they will be relatively quiet (under 2 sones) and energy efficient. Energy Star evaluates bath fans by comparing their airflow (cfm) per power consumption (watts). To be Energy Star rated a bath fan needs to be able to deliver at least 1.8 cfm/watt. In choosing the replacement fan and knowing how much air needs to be moved, you can compare the sound level and cfm/watt of the fans that meet your certified flow rate.

REPLACING AN EXISTING RANGE HOOD

If the existing range hood doesn't vent to the outside but just recirculates the air, it is the same as starting from scratch. A path to the outside is required, either straight out through the wall, up through the ceiling, or down through the floor.

If the existing range hood already vents to the outside, there are a number of options. The hood itself can be left in place and an exterior mounted, in-line fan can be installed on the outside of the building. The fan motor in the hood should be disabled so that it is not working against the new fan, and the existing grease filters, which will protect the fan motor, should be replaced with new ones.

An in-line fan in the ducting can also be used with the existing hood and filters. And to keep the sound level to a minimum, a "muffler" can be added to the duct line.

The ducting should be carefully examined, and if necessary, cleaned, before pursuing either of these options. As in any duct installation, the path should be as short and smooth as possible. If there is a damper in the exterior hood, it should be checked to make sure that it will open and close easily and seals effectively when it is closed.

Unless the kitchen is being remodeled, it is likely that there will be limitations on the size of a replacement hood. Energy Star rates range hoods up to 500 cfm, requiring an efficiency of 2.8 cfm/watt and a sound level of no more than 2 sones. The performance must be third party certified by an organization like HVI. Refer to Chapter 3 for sizing, but the hood should exhaust no more than 100 cfm per linear foot of cook-top if it is located against a wall or 150 cfm per linear foot of cook-top if it is an island installation.

INSTALLING A RADON/SOIL GAS MITIGATION SYSTEM

Radon/soil gas mitigation systems need to be sized by the needs of the specific application. If the system piping has been installed during construction, it will be relatively easy to add a fan to the system (see Figure 6.16). In most cases a 100 cfm, in-line fan can get the job done. It would be advisable to have your home checked for pollutant levels and to communicate with a certified radon mitigation contractor for the right system.

The Environmental Protection Agency (EPA) provides excellent information on how to install a radon mitigation system.

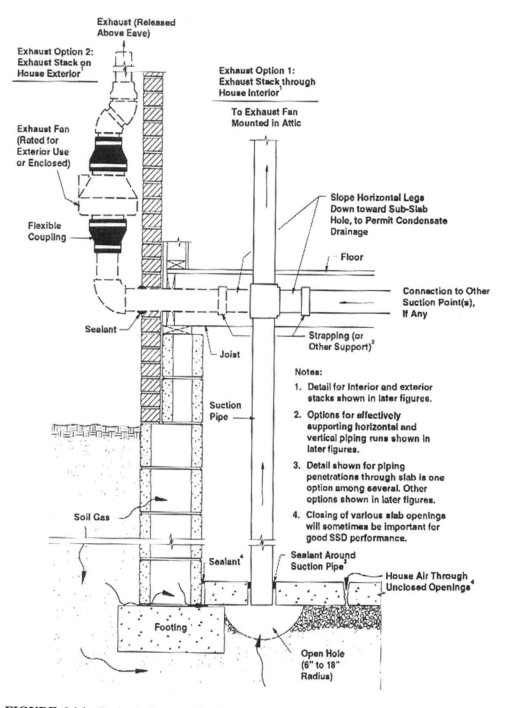

FIGURE 6.16 Radon/soil gas mitigation system. (EPA)

Chapter 7
HOUSE PRESSURES

BUILDING PRESSURES

The role pressure plays in the movement of air throughout a house cannot be overemphasized.

> **Ventilation Rule #4:** For air to move there have to be three components: air, a hole, and a force.

Since this entire book is about air movement, the presence of air is a given. Holes and forces can take all sorts of different shapes and features. Holes could be as simple as the gap around the end of a window or a poorly connected, large return register in a hallway. Forces could be mechanical devices like fans or simple variations in pressure like the wind pushing on the side of the house. Where the holes are and how big they are makes a major difference in how air much air moves through them. The strength of the force—the height of the building or the size of the fan—also has an impact on the amount of air that can be moved.

Understanding the role pressure plays in moving air around a house will provide an understanding of one of the largest components of comfort and of energy consumption.

> **Ventilation Rule #5:** High pressure always moves toward lower pressure.

Think of a balloon. Make a hole in it, and the pressurized air comes rushing out. The pressures are seeking equilibrium. When the wind blows on one

FIGURE 7.1   Wind pressures. (Panasonic/Morrissey)

side of a house, it puts that side of the house under a higher pressure. The opposite side of the house will be at a lower pressure. The air will be pushed in through holes on the high-pressure side of the house and be sucked out through holes on the low-pressure side of the house (see Figure 7.1). Air flows into a powered vacuum cleaner because the fan in the vacuum puts the end of the hose under low or negative pressure. Air is pushed out of the supply registers by the high pressure created by the supply side of the blower. On the other side of the blower the pressure is lower, and air is sucked back into the system through the return vents.

Air is constantly moving due to pressure differences in small, convective flows in wall cavities and large flows through heating and cooling and ventilation equipment, and it all has to do with pressure and holes.

STACK EFFECT

Warm air rises. Warm air is lighter than cold air and so it rises. On a still day, with no mechanical equipment running and the house out in the middle of a plain, the pressure will be higher at the top of the house than at the bottom. Holes located near the bottom of the house let air in. Holes near the top let air out. And holes in the middle have no air moving through them in either direction. These middle holes would be located in the "neutral pressure plain." This distribution of pressures is known as "the stack effect" (see Figure 7.2). A chimney relies on the stack effect to draw combustion gases in at the bottom, transport them up, and exhaust them out of the top. Until the chimney warms up, the effect is weak, which is why fires smoke until the temperature differences have been established. Once the draft has been established, the smoke (and even the fire) can roar up the chimney.

The "stack effect" is generated from the difference in weight of the warm air column within the building and the cooler air outside. The stack effect can be calculated with the following formula:

$$Q = 9.4 \times A \times \sqrt{(h \times (t - t_0))}$$

where

Q = airflow in cubic feet per minute

A = free area of inlets or outlets (assumed equal) in square feet

h = height from inlets to outlets in feet

t = average temperature of indoor air at height h in °F

t_0 = temperature of outdoor air in °F

9.4 = constant of proportionality, including a value of 65 percent for effectiveness of openings. (This should be reduced to 50 percent [constant = 7.2] if conditions are not favorable.)[1]

So if the house is 18 feet tall and there is a 1 foot hole at the top and a 1 foot hole at the bottom and it is 70°F inside and 45°F outside, an airflow of about 200 cfm will be generated without a fan!

When the wind blows against the side of the house, the design of the house has an impact on the shape of the flow. For a simple shape, the side of the

1. Brumbaugh, James E., Audel HVAC Fundamentals Volume 3, Wiley Publishing, 2004.

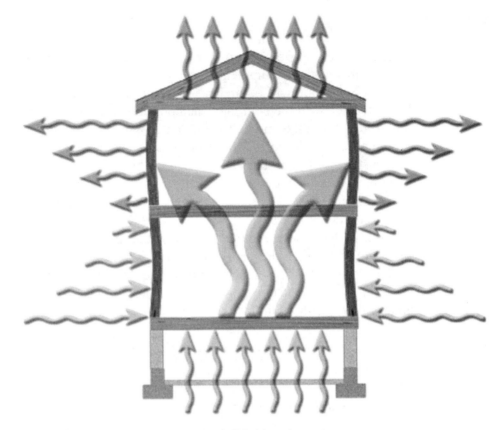

FIGURE 7.2   Stack effect. (Panasonic/Morrissey)

house that the airflow impacts, the pressure rises. The air stream moves up and over the roof and down the other side, creating a negative pressure or vacuum condition on the back and sides of the house as the air moves away. These pressure conditions impact the flows due to stack effect in the house.

The wind is a variable force. Averaging the winds over the course of the year may indicate that there is adequate natural airflow to eliminate the need for mechanical ventilation. But the need for ventilation in the home is constant. In the seasons when the heating system isn't running or the windows are closed or the temperature inside is close to the temperature outside, the natural air movement forces just can't get the job done.

Ventilation Rule #6: A constant need cannot be satisfied by a variable, average force.

DIFFUSION

A high concentration of a gas will move toward a lower concentration, seeking to balance out the two, seeking equilibrium. This is the process of diffusion or vapor pressure. Cooking bacon in the kitchen will eventually diffuse the fragrant molecules throughout the house. The room deodorants or air "fresheners" move their fragrances throughout the room by diffusion as well as airflows. Gases and water vapor can diffuse through solid materials like walls, moving slowly from molecule to molecule.

Diffusion is a much slower process than the movement of air mechanically or naturally with the other pressure forces of winds and stack effect. It is a force that should be recognized and appreciated, but on the scale of air movement in the house, diffusion is not a major ventilation factor.

LEAKY DUCTS AND AIR HANDLING EQUIPMENT

Leaky ducts, on the other hand, can have a major impact on the movement of air in the house. The heating or cooling system that uses conditioned air is meant to be a closed system. House air moves into the return or low-pressure side of the air handler, and is pushed out through the ducts into the house on the high-pressure side. It moves through the rooms and back into the returns and back to the air handler. The same amount of air that is pushed out is drawn in.

If the ducts are tightly connected (all the joints sealed with mastic for minimal leakage) or if they are all located within the conditioned space, the system should work as designed. But if there are leaky ducts that are located outside the conditioned space, there will be an impact on the pressures in the house that will drive airflows.

The air handler doesn't care where it gets its air from, it will continue to push air out and suck air back in regardless of the ducting. Leaks in ducting on the return side, drawing a portion of their air in from outside of the conditioned space, put the conditioned space under positive pressure because they take less air back to the air handler than the system is putting into the conditioned space.

Leaks in the ducting on the supply side can have the opposite effect. If the supply ducting leaks to the outside, it puts the building under negative pressure because the air handler will be adding less air to the conditioned space than it is taking out through the return ducts. Of course at the same time the leaky supply ducting is conditioning the outdoor world.

In a cold or heating system dominated climate, putting the house under positive pressure, even if it is very slight, will force the warm, potentially moist air in the house to leak out through any cracks or gaps or holes that it can find. When that warm, moist air strikes a cold surface below the dew point, the moisture in the air will condense out. That cold surface may be somewhere in the wall or roof system of the house. That moisture will encourage the growth of mold and degradation of the building materials. The same effect will occur in the opposite direction in a cooling climate when the house is under negative pressure from leaky supply ducting. The warm, moist air will be sucked in from the outside. These air leaks and the associated flows may be very small, but small "drips" can accumulate. Nature is very patient.

It would be highly unusual for the duct installation to be completely air tight on only one side of the system. It is likely that there will be pressure leaks on both sides. It's more a question of which side has more leaks. It is difficult to determine that, and it doesn't really matter. It is more important to seal all the duct joints or install all the ductwork within the conditioned envelope.

EFFECTS OF PRESSURE ON BACKDRAFTING

Chimneys are designed to guide polluted, unwanted air, gas, smoke, and particulates from combustion appliances to the outside of the house. They work by carrying the pollutants in the air stream. The chimney is the "hole." The "force" is assumed to be there by the "stack" effect of the warm, rising column of air. However, just because there is a chimney or vent pipe doesn't guarantee that the air will move up and out through it.

The venting system should be designed and installed according to local building code. All the factors like the length of the horizontal run, the number of elbows, and the height of the vertical rise should be figured in. (Notice the similarity in other ducting resistance issues?) The National Fuel Gas Code (NFGC) or ANSI Z223.1 or NFPA 54 sets the venting requirements for all natural draft, Category 1 gas-fired appliances. The Category 1 vent systems work by allowing the heated pollutants to rise vertically. The longer the horizontal run relative to the vertical rise, the more work the pressure has to do to get the air to flow up and out through the vent.

A Pascal is a tiny amount of pressure something akin to a dust mite's sneeze. It is important to recognize the major effects such a seemingly minor force can have on an object as large as a house.

Note that a gas vent is not sealed to the gas appliance (unless it is a "sealed combustion system"). There is an opening between the vent and the boiler or furnace that is protected by a "flue hat." As the exhaust air moves out of the combustion appliance, room air is drawn in around the edges of the "hat." Air can also be exhausted out of the vent and appliance into the room if the air pressure in the room is lower than the air pressure in the vent: that pressure can be as little as 3 to 6 Pascals—a very, very small amount. For the health and safety of the house's occupants, the appliance venting must be installed to code and the manufacturer's instructions, and there must be an adequate pressure difference (higher in the space and lower in the vent pipe) to compel the combustion gases to move up the chimney and out of the building.

Using a sealed combustion system, a system that mechanically draws its combustion air in from the outside and power vents the flue gases back to the outside independently from the house air, assures that pressure variations in the house will not impact the performance of the system. Power assisted combustion appliances, appliances that use a fan on the exhaust side, are the next best step to ensure adequate venting.

Of all the pressure issues in a house, venting of combustion appliances is the most immediate and critical. Small changes have a big impact (see Figure 7.3). The effects of very low-level carbon monoxide poisoning are still being studied. The effects of moderate levels of carbon monoxide poisoning can be deadly.

ATTACHED GARAGES

Garage doors are large holes. Even when they are closed, most garage doors are not well sealed around the edges. And they are located at the lowest level in the house. (It's difficult to put the garage in the middle of the building!) As such, most of the time the air pressure is pushing in on the doors and other leakage paths. Unfortunately they are usually just like another room in the house, disconnected from the outside by a door, but not well air sealed from the house. So if the pressure in the garage is higher than the pressure in the house, all the pollutants that are in the garage will flow into the house.

This may be corrected by adding an exhaust fan to the garage (the International Mechanical Code says 100 cfm per car[2]), but if there is a combustion appliance in the garage (like a water heater), an exhaust fan may cause backdrafting as mentioned above.

2. IMC, Table 403.4.

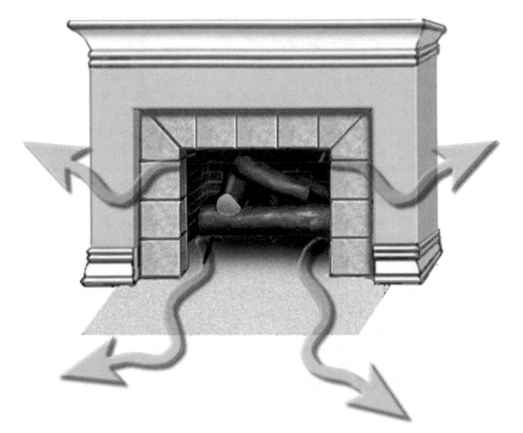

FIGURE 7.3 Fireplace backdrafting. (Panasonic/Morrissey)

If there is ducting in a garage, it is particularly important that all the joints be air sealed with mastic, otherwise those holes can suck in all the pollutants in the garage and distribute them throughout the house. ASHRAE's ventilation standard (62.2-2007) allows "no more than 6 percent leakage of the total fan flow when measured at 0.1 in. w.g. (25 Pa)"[3] in the garage.

If the air handler itself is in the garage, it is almost a sure thing that the garage pollutants will be distributed to the house. The appliance manufacturers will not allow the air handler's housing to be completely air sealed. The air handler could be boxed off from the garage and provided with an access door to the exterior to allow for service. The joints in the walls of the air handler closet must be carefully air sealed.

3. ASHRAE Standard 62.2-2007, p. 6.

Ideally, the garage should be physically separated from the house. If that is not the case, keeping the mechanical equipment including the ducting out of the garage, air sealing the barrier between the house and the garage, and installing an exhaust fan in the garage is the next best step. If the combustion appliance has to be in the garage, using sealed combustion eliminates the problem of venting the garage and putting it under negative pressure.

HOUSE TIGHTNESS OR BUILDING TIGHTNESS LIMITS (BTL)

Can a house be too tight? If the house will not allow the occupants to live comfortable and healthy lives and pursue their personal activities, yes it can. The house is a place to shelter and protect the occupants. It must allow them to live comfortably.

Building tightness limit (BTL) (or building airflow standard [BAS] or minimum ventilation level [MVL] or minimum ventilation rate [MVR]) is a gen-

---

The BTL can be calculated using three formulas:

Formula #1: 15 cfm × the number of occupants × n = CFM50 BTL
This formula estimates a tightness level based on the number of occupants.

Formula #2: 15 cfm × number of bedrooms + 15 × n = CFM50 BTL
This formula calculates the tightness level based on the number of occupants as estimated by the number of bedrooms.

Formula #3: Volume of the conditioned space × .35 x n/60 = CFM50 BTL
This formula calculates the tightness level based on .35 air changes per hour of the volume of the building.

The BTL equals the highest number calculated using formulas 1, 2, or 3. The number recorded would be the minimum allowable CFM50 of the conditioned living space. For example a one-story, 1200 square foot, two-bedroom house in Massachusetts with three occupants is tested with a blower door to 960 CFM50. (For this house in this location n = 14.8 from the LBNL tables.)

Formula #1: 15 cfm × 3 × 14.8 = 666 CFM50 BTL

Formula #2: 15 cfm × 2 + 15 × 14.8 = 666 CFM50 BTL

Formula #3: 1200 × 8 (ceiling height) × .35 × 14.8/60 = 828 CFM50 BTL

All of these calculations are less than 960, indicating that this house would be loose enough not to require mechanical ventilation. A better alternative would be to tighten up the house further, dropping it below 650 CFM50 and then adding mechanical ventilation.

eral term for a house-tightening limit used for ensuring adequate air quality for the occupants of the house.[4] Weatherizing or reducing the energy consumption of a house requires tightening it up, closing up the air leaks and holes where pressure forces the air in and out of the building. Many homes, particularly manufactured homes, get a large portion of their air from the air leaks, and once those have been carefully sealed, the requirement for mechanical ventilation increases because the natural air changes have decreased. The building tightness limit was developed as a benchmark, a limit beyond which the health of the occupants may be jeopardized. Most homes can use a primary ventilation system. The BTL benchmark should be considered a way to confirm what ventilation is needed.

The tightness of the house can be measured using instrumentation such as a "blower door." Many green building and weatherization programs have criteria for how tight a home should be before mechanical ventilation must be added. This may be a reference to a BTL or to a minimum ventilation rate (MVR). Indiana, for example, sets its MVR at 1200 cfm@50 Pa. (This is the "CFM_{50}" level.) That means that a blower door must be used to exhaust air from the house until the house reaches a negative pressure relative to the outside (with respect to or WRT) of 50 Pascals. If the amount of air flowing through the test fan to accomplish this exceeds 1200 cfm, then the house is leaky enough not to need additional mechanical ventilation (under its program). Luckily homes are not being continuously exhausted to this artificially high level, which is used to exaggerate the conditions and to reach a standardized, comparative level. The CFM_{50} level can be converted to a more realistic $CFM_{Natural}$ or CFM_{Nat} level by dividing it by a factor that takes into account the location or wind conditions and the height of the building known as the "N" or "n" factor. These vary from a low of 9.8 for an exposed three-story house in a cold climate like North Dakota to 29.4 for a well shielded, one story house in a warm climate like Miami.[5] Dividing the CFM_{50} number by the "n" factor provides CFM_{Nat} or approximately how the air will move through the home under average natural conditions.

For example, a two-story 1200 CFM_{50} house located in Massachusetts that would be considered "normal" in terms of wind exposure would use an "n" factor of 14.8. Its CFM_{Nat} would be 1200/14.8 or 81 meaning that under average conditions this house would be leaking 81 cubic feet of air per minute. If

4. Survey of Tightness Limits for Residential Buildings, July 2001, Rick Karg for the
    Chicago Regional Diagnostics Working Group, available at www.karg.com/btlsurvey.htm.
5. Krigger, Dorsi, <u>Residential Energy</u>, Appendix A-11, p. 284.

the house were tightened up to 900 CFM_{50}, its CFM_{Nat} would be reduced to 900/14.8 or 60 cubic feet per minute. This is under "average" conditions. Much of the time it would be leaking less than this.

Because it takes such a small amount of pressure to potentially backdraft a combustion appliance, caution should be taken in aggressively tightening up the house and employing mechanical ventilation. It is highly recommended that combustion safety testing be done before, during, and after the air sealing/tightening process. The calculated CFM_{Nat} number is an approximation under "average" conditions, meaning that half the time the natural ventilation will be greater than this and half the time it will be less than this. Using a continuous, primary mechanical ventilation system assures that some ventilation will be happening all the time and is an advisable approach unless the house is obviously too leaky to warrant it.

Chapter 8
PASSIVE INLETS, OUTLETS, TRANSFER GRILLES, AND MAKE-UP AIR

A wise man once said, "People are systems. They have specific openings for specific purposes. Air for breathing goes in and out of specific, controlled openings, not randomly all over the body." Houses are systems and they should have specific openings for specific purposes as well.

Ventilation means exchanging air between the inside and the outside of the house and circulating it around the living spaces. Mechanical equipment only provides part of the solution. Exhaust fans, range hoods, central vacuums, and clothes dryers mechanically exhaust from the house. Fireplaces, water heaters and furnaces passively exhaust air and pollutants from the house through their chimneys. It is common practice to consider that there will be more than enough leaks and holes in the house to provide an adequate amount of "make-up" air. As homes become tighter that is not always true (see Figure 8.1).

Getting air in from the outside to the inside and back out to the outside is only part the challenge. Circulation, moving the air throughout the living spaces, is required to keep the building and the occupants healthy. Ideally every room would have a correctly placed supply and return for heating and cooling and a second pair for ventilation. That way adequate circulation could be assured. However, since the object is to circulate fresh air through the rooms, using the rooms as the conduit is a reasonable approach. After all, people don't live ducting.

Rooms have doors, however, and doors serve as dampers, manually operated blockages to airflow. Doors that are sealed well enough to block the transfer of sound and provide privacy also block the flow of air. The crack under the door is usually only enough to allow the bottom of the door to clear the carpeting and open and close properly. That crack is often relied upon for the transfer of conditioned air, balancing the pressure between the room and the rest of the house. To do that effectively the crack needs to be more than a crack.

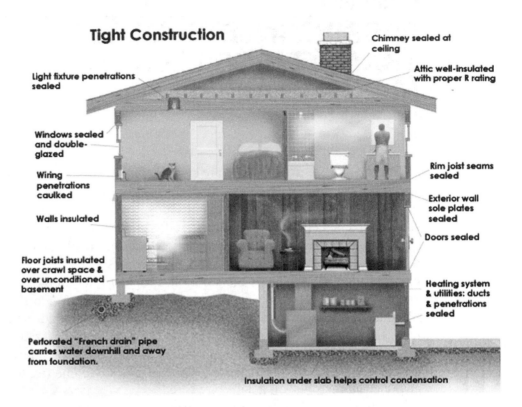

Tight Construction

Chimney sealed at ceiling

Attic well-insulated with proper R rating

Light fixture penetrations sealed

Windows sealed and double-glazed

Wiring penetrations caulked

Walls insulated

Rim joist seams sealed

Exterior wall sole plates sealed

Doors sealed

Floor joists insulated over crawl space & over unconditioned basement

Heating system & utilities: ducts & penetrations sealed

Perforated "French drain" pipe carries water downhill and away from foundation.

Insulation under slab helps control condensation

FIGURE 8.1   An example of tight construction. (Panasonic/Morrissey)

OVER, UNDER, AND THROUGH-WALL CIRCULATION

Just like the ducting design, the circulation issues for ventilation are different than the circulation issues for the HVAC system. The airflows and pressures created by the HVAC system are considerably greater than for ventilation alone. If ventilation has been integrated with the HVAC system, then, of course, they have become one and the same. But if the ventilation strategy is to use a separately ducted, positive pressure system, a non-integrated HRV or ERV, or some other dedicated ventilation ducted approach, circulation solutions can be different. If the HVAC system has a dedicated supply and return in each room, then bypassing the closed door may be limited to ventilation issues. But, as is typical of most HVAC systems, each room has a supply and the return is in the hallway, then the bypass must serve for both ventilation and conditioned air.

For ventilation alone, the opening between the hall and the room can be fairly small. If the room is being supplied with 30 cubic feet per minute or less, the typical half-inch crack under the door will suffice. A 30-inch wide door with a ½ inch undercut would provide a 15 square inch opening, which would be adequate to keep the pressure in the room below 2 Pascals relative to the rest of the house. A door insert (a grille that mounts in the door and blocks the transfer of light or sound) works for just ventilation air pressure relief. The resistance of the grille in a door insert, however, offers much too much resistance for HVAC pressure relief, and since both ventilation air and conditioned air must be circulated, the higher-pressure requirement of the conditioned air must be used to determine the dimensions of the pressure relief.

Generally, both conditioned air and ventilation air will need to bypass the door. The object is to allow the air to flow from the room under, through, over, or around the closed door and keep the pressure from one side of the door to the other at 2.5 Pascals or less. Ideally there would be no pressure differential, but that is difficult to achieve.

To go over the door, from one side of the wall to the other, a "jumper" duct can be used (see Figure 8.2). A jumper duct is simply a piece of ducting that is attached to a "collection box" or boot in the ceiling covered by a grille on one side of the wall to a box in the ceiling also covered by a grille on the other side of the wall. This allows the air to flow, pushed by the pressure imbalance from one side to the other. It also reduces the flow of light and sound, affording some privacy. The simplest approach is to use flexible ducting for this, which, with its rough interior surface, is resistive to easy airflow. The grilles are also barriers to the small pressures referred to here. If just 30 cfm of ventilation air is being delivered to the room, the jumper duct could be constructed of 6-inch diameter flexible ducting. If 100 cfm of conditioned air is being delivered to the room as well as 30 cfm of ventilation air, the duct diameter should increase to 12 inches. Tests at the Florida Solar Energy Center have shown that the diameter of the jumper duct should be approximately equivalent to the square root of the delivered airflow to keep the pressure differential below 2.5 Pascals. This takes into account the other components of the system including the duct boxes and the grilles.

For flexible "jumper" duct: $diameter = \sqrt{airflow}$ **(delivered to the room)**

The slot under the door is simple and doesn't require concern about grille resistance (see Figure 8.3). A 1-inch slot under a 30-inch wide door will allow approximately 60 cfm to be delivered to the room and maintain the 2.5 Pascal pressure difference. (The same FSEC tests showed that doubling the area of

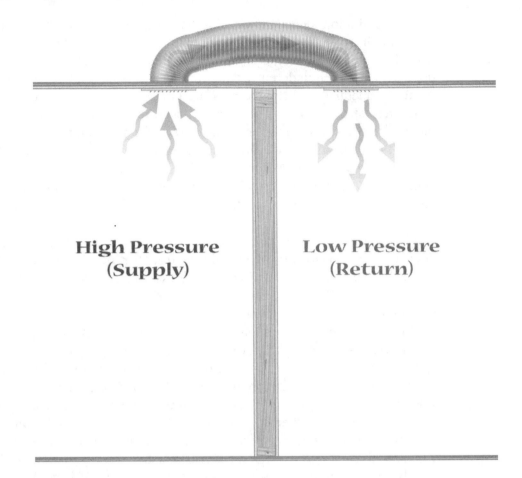

**High Pressure
(Supply)**

**Low Pressure
(Return)**

FIGURE 8.2   A jumper duct. (Morrisey)

the slot would allow for doubling the airflow.) For 130 cfm the slot would have to be cut to approximately 2 inches. A slot that large will certainly infringe on privacy and reduce the purpose of closing the door in the first place.

For slot under door: $Area = \frac{1}{2} \times airflow$ **(delivered to the room)**

Going straight through the wall is another approach that provides limited privacy (see Figure 8.4). It is also important for through-wall approaches to have a sleeve that completes the passage from one wall surface to the other, otherwise air from the wall cavity will be drawn into the system. The majority of the air resistance for the straight-through-the-wall approach will

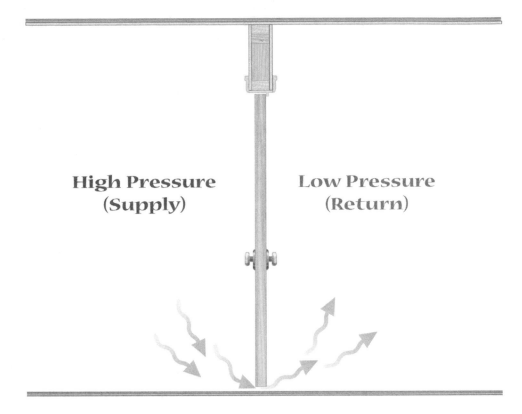

FIGURE 8.3   A door undercut. (Morrissey)

come from the grilles covering the opening. Dividing the airflow by .83 provides the area of the opening. Using 130 cfm again, the opening would need to be 156 square inches, perhaps 10 inches by 15 inches or 10 inches by 16 inches, a large opening.

For grille covered opening through-wall: $Area = airflow/.83$

Offset grilles are an approach that does afford some privacy since the hole does not go straight through the wall (see Figure 8.5). The flow, however, will draw cavity air into the system and the maximum flow to keep the pressures below 2.5 Pascals is limited to 72 cfm if it is a standard 2 x 4 constructed wall. No matter how big the openings are on either side of the wall, the limiting factor for the flow is the thickness of the cavity. In fact, increasing the openings beyond 112 square inches has negligible impact on the flows in an offset grille arrangement.

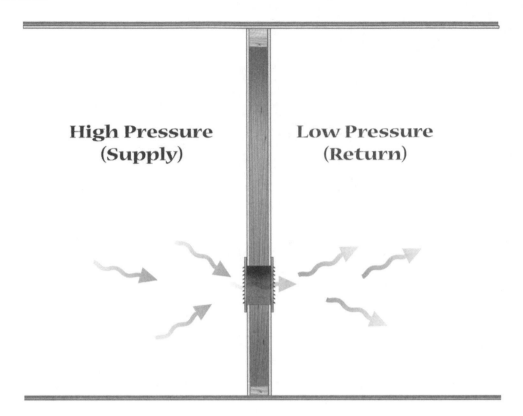

FIGURE 8.4 Straight through the wall pressure relief. (Morrissey)

Of course these transfer methods are additive. It is likely that there will be some space under the door in any case. That opening can reduce the size of other approaches that are used. And as stated previously, the smaller airflows and lower pressures involved with ventilation mean smaller relief "holes" can be employed IF ventilation circulation is the only concern.

INLETS AND OUTLETS

Okay. So you go to all this effort to seal up every crack, hole, gap, and chink in the envelope of the house and now you need to intentionally makes holes in that envelope to let air in. Why not just save all that sealing up time and effort and leave the existing holes? Mainly because all the inadvertent holes and gaps are in uncontrollable places, many in the wrong places, and there are too many of them adding up to a much larger "hole" than is needed for ventilation air.

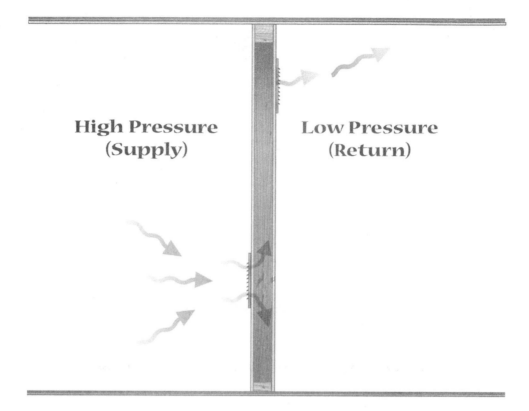

**High Pressure
(Supply)**

**Low Pressure
(Return)**

FIGURE 8.5   Offset grilles. (Morrissey)

What you need are affectionately called "smart holes," trickle vents, or devices that are designed to allow a specific amount of air to pass through them no matter what the weather conditions (see Figures 8.6 and 8.7). They often include filters, which will remove the large particles that might be floating around in the outside air.

The performance of trickle vents or passive inlets is affected by unintentional building leakage to the exterior as well as between living units, the ventilation rates, the control of heating and cooling in the bedrooms, the acoustic performance requirements to minimize outside noise infiltration, the location of the inlet to minimize the perception of drafts, and the design of the inlet itself. Trickle vents and inlets are not stand-alone components and won't function effectively without the remainder of the system, particularly in relation to resolving indoor air quality problems.

Bear in mind that these small holes are not designed as make-up air sources for major air sucking appliances and other devices like clothes dryers,

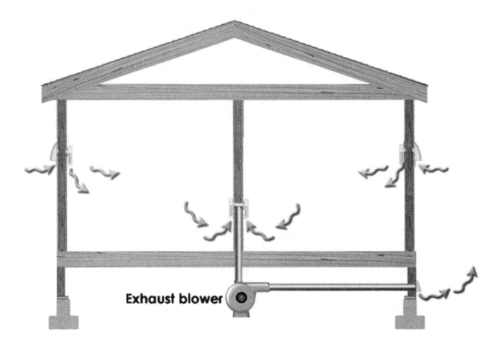

FIGURE 8.6 Exhaust-only with inlets. (Panasonic/Morrissey)

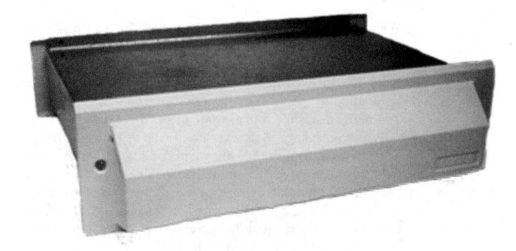

FIGURE 8.7 Fresh air inlet. (American Aldes)

range hoods, fireplaces, or woodstoves. If those devices are activated, they will draw air from any place that is an opening to the outside, but these "smart" holes should not be their primary source of air.

Trickle Vent/Smart Hole Design

Trickle vents are not simply holes with grilles on either side. They are fairly complex products that have been used for many years in Europe and other parts of the world (see Figure 8.8).

Through-the-wall "passive inlets" generally include an exterior hood to keep the weather out with critter and insect protection, a through-wall fitting or short duct length that is either round or rectangular, an air filter, possibly sound abatement padding, and an interior cover that sometimes includes a means to close the system completely. Some of them have humidity-controlled dampers that shut down as the level of relative humidity increases. The systems are designed to regulate the amount of air that is able to pass through them which can be as low as 10 cfm to not more than 20 cfm.

For trickle vents to work effectively as inlets, the air pressure in the room must be lower than the air pressure outside. The building needs to be tight so that the vents comprise the majority of the building leakage. When the building is leaky, the inlets have to be made larger to assure that the airflow through the dedicated inlet supplies the majority of the air to the room. If the inlets are made too large, the building is much more subject to the vagaries of the wind

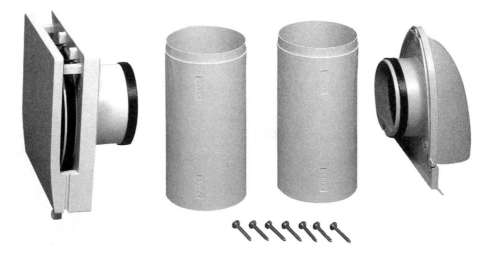

FIGURE 8.8   Inlet vent components. (Panasonic)

and stack effect. This becomes less of a problem if the vents are not intended to be associated with a mechanical ventilation system but are simply a means for adding additional leakage at the windows, allowing the air to flow either way across the openings (see Figure 8.9).

Location

Where the "smart holes" are located will greatly affect how they perform as inlets. Just because a "hole" is supposed to serve as an inlet doesn't mean it will work that way. It will depend on the building pressures described in Chapter 7. These devices are smarter than using a hammer to smash an opening in the wall, but they're still just holes. The laws of stack effect and wind loads still apply.

Ideally the trickle vents are located in bedrooms where people spend the most time. Air is drawn in from the outside, flowing through the bedroom to the primary ventilation system located near the center of the house. If the

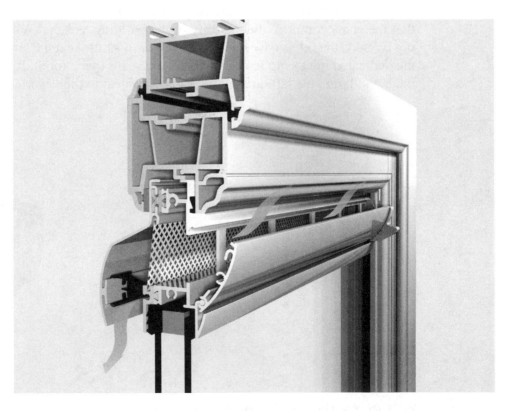

FIGURE 8.9   A "trickle vent" in a window. (Titon)

"holes" are located high on the walls, the incoming, untempered air enters at a very low flow rate that is barely perceptible by the occupants. It mixes with the room air and enters above their heads. On a single story home, however, openings in the walls near the ceiling may serve as outlets because of the stack effect.

If the inlets are located on the side of the building facing the prevailing wind, they are more likely to serve their purpose. Air pressing on that side of the house will be forced through the vent, and the internal mechanisms will limit the amount of air that enters.

When trickle vents allow the air to flow out of the building, it is not a disastrous situation since the amount of air that moves through them is limited. It is just another air leak, and it won't be providing the fresh, make-up air that the design requires.

DESIGN AND TEMPERING THE MAKE-UP AIR

When introducing outside air into the living space, the temperatures will not be the same. When the outside air is warm and humid, it will add to the load of an air conditioning system, but tempering is less critical for comfort than when the outside air is frigid. Make-up air that is added directly to the return side of an air handler will be tempered as it passes through the conditioning elements.

Small amounts of make-up air (20 cfm or less for bedrooms, for example), introduced into the living space at a low velocity and out of the occupied space, will blend with the room air and cold drafts will be avoided.

If a large amount of make-up air is required (for range hoods, for example), fan forced tempering will definitely be needed.[1] For commercial applications, a number of companies manufacture make-up air units, particularly for kitchens. These can have heating or cooling components to temper the thousands of cubic feet per minute flowing back into the building. Table 8.1 describes the amount of electric heat required to raise the outdoor temperature for different airflows at different temperature differentials. For example, if it is 20°F outside and you want to raise the temperature to 60°F, 1000 watts is needed for an 80 cfm system. A 400 cfm would require 5000 watts (5kW).

An air duct bringing combustion air to a fireplace is generally not fan forced, relying solely on the stack effect to draw air through the ducting and

1. There are a number of companies that produce tempered make-up air systems and a list can be found in Appendix F.

TABLE 8.1 Electric heat for conditioning make-up air.[2]

Watts	50 degree difference	40 degree difference	30 degree difference
1000	60 cfm	80 cfm	100 cfm
2500	160 cfm	200 cfm	260 cfm
5000	320 cfm	400 cfm	525 cfm
10,000	630 cfm	785 cfm	1050 cfm
15,000	945 cfm	1185 cfm	1580 cfm

2. Adapted from http://www.electromn.com/gen/makeup_air.htm.

through the combustion chamber and up the chimney. When the fire is burning, tempering of the incoming air is occurring. When the fire is not burning, that duct is a conduit from the outside; air will leak into it and out of it depending on the pressures in the house. The flows will generally be small enough so as to be unnoticeable. (But in terms of energy efficiency of the building, these make-up air ducts can be a significant part of the total leakage component of the house.)

Clothes dryers draw a significant amount of air out of the house (commonly between 100 and 200 cfm). Gas dryers also need air for combustion. A make-up air vent in the vicinity of the dryer (even underneath it) can improve the efficiency of the drying cycle. A pressure switch controlling a motorized damper can close off that intake line when the dryer is not operating.

HRV/ERVs as primary ventilation systems temper their own air, but they should not be relied on to provide extra make-up air to the living space. They are designed as "balanced" systems, bringing in the same amount of air as they exhaust. They will not provide "extra" air for other appliances such as range hoods or fireplaces.

Chapter 9
VERIFICATION OF PERFORMANCE AND TESTING

FLOW TESTING PROCESSES AND EQUIPMENT

The only way to be sure an installed system is operating the way it was conceived and designed on paper or by computer is to test it. Anything else is just a guess. All of the estimates of equivalent duct length and laboratory testing are just estimates and only approximate the actual installed conditions. The fact is that in a real house construction elements are not always in exactly the spots that they were called out to be on the plans (if there are plans). There will inevitably be places where the ducting has to make additional twists and turns to get to the outside of the building. There will be places where someone just may have stepped on a piece of flexible ducting in the attic or was not quite able to make a connection air tight.

The most important element to test in any ventilation system is the airflow. Because residential flow testing is not a requirement in the United States, the available equipment has been adapted from commercial equipment, designed for higher airflow rates than are seen in most residential applications. Flows in homes as low as 30 cfm are difficult to measure partially because the flow measuring device has to be in the air stream and anything that gets in the way of the airflow will affect the airflow.

Processes can be as simple as holding up a hand and feeling the flow of air or the use of "particle streak-velocimetry" or tracer gas to actually "see" the movement of air. In measuring the performance of a ventilation system, we are generally concerned with the flow of air through the grille or fan that is located in the room. Flow measurements can certainly be made in the ducting and should be for heat or energy recovery ventilators in order to balance them. Measuring the flow in the ducting requires making a hole for the probe or adding a small section of ducting in which to install a "flow grid."

Pieces of paper—Paper works as a gross indicator of airflow through an exhaust fan (see Table 9.1). As the air flows up to the fan opening it can draw paper up to the grille. If it is strong enough, it will actually hold the paper over

TABLE 9.1   Fundamental airflow management.

Material	Flow that draws it up to the grille
Single ply toilet paper	> 30 cfm
Double ply toilet papers	> 60 cfm
"20 pound" copier paper	> 110 cfm

the grille opening. Different weights of paper can indicate different airflows. At least you'll know if air is moving through the fan. You won't know if it is flowing out of the building.

Note that this is a gross approach, only slightly more accurate than a "calibrated hand," and should not be used for anything more than a relative measurement of airflow. However, if the fan can't lift any of these papers, it is pretty clear that something needs to be done to improve the flow.

Garbage bags[1]—The Canadian Mortgage and Housing Corporation (CMHC) determined that air moving out or evacuating a space could deflate a pliable bag, like a garbage bag (for example a "Glad" 66 x 91 cm). The amount of time it takes to accomplish that is parallel to the amount of air that the ventilation device is moving out of the space. The opposite is true for a supply ventilation system, where the air will inflate the bag. Tape the opening of the garbage bag to an expanded wire hanger, scoop up the air, hold the opening over the grille of the fan, and time the deflation. The deflation time is an indicator of the airflow rate: 1 second = 50 cfm, 5 seconds = 30 cfm, 12 – 13 seconds = 10 cfm. Although this approach sounds crude, it is surprisingly accurate and certainly inexpensive.

Hot-wire anemometers—When air moves across a heated wire (commonly tungsten), the wire seeks to maintain its temperature as the air moves across it at a rate that is relative to the airflow. The heated wire is extremely small so the flow is a measurement at that particular point (see Figures 9.1 and 9.2). To get the total flow across a fan opening or duct, multiple measurements need to be made and averaged out. The result is a measurement of the velocity of airflow (feet per minute) not volume of airflow (cubic feet per minute), which would be calculated from the average velocity multiplied by the area of the opening. The area of the opening must be calculated carefully, removing the area blocked by the ribs of a grille, for example, which can be as much as

1. http://www.cmhc-schl.gc.ca/en/co/maho/yohoyohe/inaiqu/inaiqu_003.cfm.

50 percent of the opening. Some of these devices can accept a vent area and calculate the volume of flow. Hot wire anemometers can sense flows from 0 to 10,000 feet per minute.

<u>Vane anemometers</u>—A vane anemometer is a small fan blade that spins when it is positioned in an airstream (see Figure 9.3). The more air that moves across it, the faster the fan spins. Mini-vane anemometers can be as small as

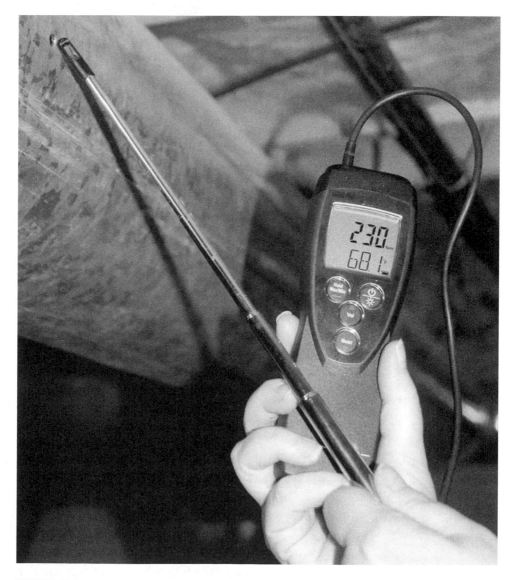

FIGURE 9.1   A hot-wire anemometer. (Testo)

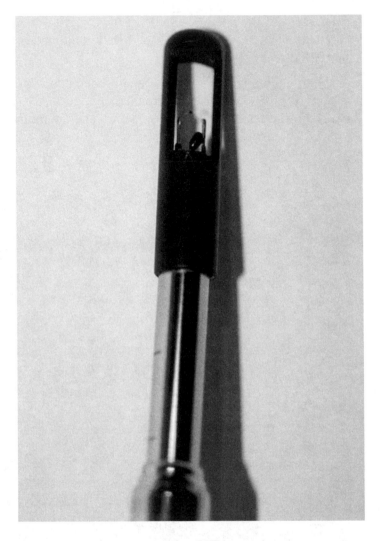

FIGURE 9.2 A hot-wire anemometer tip. (PHR)

½ inch in diameter. Measurements with these devices are similar to measurements with hot-wire anemometers, a single point. Multiple measurements need to be made and averaged and multiplied by the area of the opening to convert from feet per minute to cubic feet per minute.

Large vane anemometers that can typically measure flows from 50 to 6000 fpm can be purchased with hoods that will cover the fan opening and internally convert from velocity to volume (see Figure 9.4). They are not dependent on the

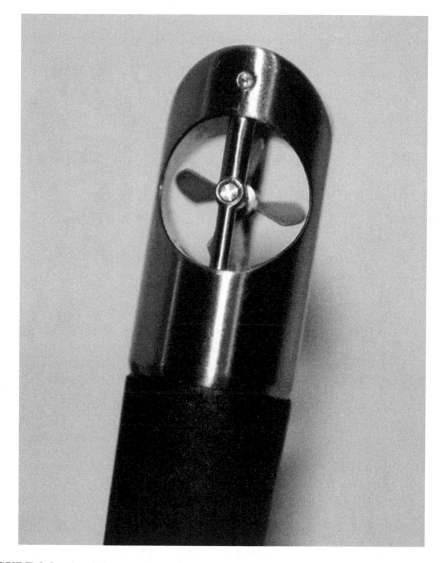

FIGURE 9.3   A mini-vane anemometer tip. (PHR)

density of the air or compensation due to air temperature, humidity, or atmospheric pressure to generate an accurate measurement. Every effort is made to make these devices as low resistant as possible to spinning in the airstream. But it takes at least some energy away from the flow being measured to make the blade of the measuring device spin. At the same time the body of measuring device is in the air stream and that also has an impact on the flow of air. When

FIGURE 9.4   A large vane anemometer. (PHR)

there is a need to measure flows as small as 5 or 10 cubic feet per minute, these impediments have more impact than in an application where the flows can vary by 100 or 200 cfm and be acceptable. The accuracy of vane anemometers is also affected by the angle they are held to the direction of the flow. At these low flows it is important to hold the device as close to perpendicular to the flow as possible so that the vanes will spin with the greatest accuracy.

Flow hoods[2]—Flow hoods are collapsible, fabric hoods which serve as a flow capture device and are large enough to cover the grille with an internal

2. Walker, Wray, et al, "Evaluation of flow hood measurements for residential register flows," LBNL, September, 2001.

flow measurement grid, hot-wire, pressure drop, plate deflection, or vane design and generally include a built-in digital meter. Small residential grilles often cause enough turbulence to reduce the accuracy of these devices. There is a wide range of issues that affect the performance of these devices on residential applications including the placement of the hood over the grille being measured. It is important to center the grille in the hood opening as much as possible.

Pressure pans or exhaust fan flow hood[3]—Standard pressure pans are primarily used to cover openings and measure pressure drops or losses through the ducting. A box with an adjustable or calibrated opening can be used to measure the flow through a fan. The airflow through a known sized hole can be calculated if the pressure differences are known using the formula:

where

$$Q = 1.07 \times A \times \sqrt{\Delta P}$$

Q = airflow in cfm

A = the area of the hole in square inches

ΔP = the difference in pressure in Pascals

Devices that are specifically made for this purpose include a fabricated box with an adjustable, calibrated opening, and an attachment for a hose that is connected to a digital manometer. They can measure flows between 10 and 125 cfm and often include a pole to hold the box up on the ceiling, covering an exhaust fan opening.

Duct testers (powered flow hood)—Duct testers are calibrated fans that can be used to pressurize a duct system to measure how leaky it is. The same system can be used to accurately measure the flow through a ventilation fan. The ducting from the calibrated fan is installed over the grille or opening to the fan being tested. The flow through the calibrated fan is increased until there is no pressure difference between the two fans. At that point, the flow through the calibrated fan will be equal to the fan under test, and the flow reading can be read directly on the digital manometer or pressure gauge. This eliminates all the effects of the testing device and is a very effective means of getting an accurate flow measurement even at very low flows.

---

3. Note that this does not refer to the "pressure pan" sold by the Energy Conservatory to measure duct leakage when making a blower door test.

INSTALLED SYSTEM TESTING AND BALANCING

Although they are known as "exhaust-only" or "supply-only" ventilation systems, that obviously doesn't work based on the fundamental principal of "one cubic foot in equals one cubic foot out." Even though the mechanical system is only on one side of the equation, the air is dragged in or forced out from somewhere to make up the balance. So if you are testing an exhaust-only system for example, and you don't get the flow you expected, you might try opening a window and testing again, (although it is far more likely that the ductwork, grilles, and the rest of the system itself is the culprit). Although it would be a remarkably tight house, you might notice a difference in airflow in which case you might have to consider adding a second mechanical system to balance things out.

For an existing ventilation system the first step is to review the installation to determine what sort of system it is—supply-only, exhaust-only, heat recovery ventilator (HRV), energy recovery ventilator (ERV)—and perform some common sense visual checks to make sure that the system has been installed in such a way that it will function as designed and as required for the application. It would be a true waste of time to do a series of performance checks only to find that the outlet vent had been stuffed full of socks! A physical inspection of the system will tell you a lot before you start measuring airflow, and you need to know how the system was supposed to work and how to turn it on and off before you can test it.

How is it ducted? Is it integrated with the HVAC system or does it stand alone? If it is an outside feed to the return side of the HVAC system, is there a motorized damper in the line to the outside? Is there a flow regulator that will allow only a defined amount of air into the system? (A feed into the return side will make the system essentially a positive pressure ventilation system.)

If it is an HRV or ERV, does it have its own fully ducted distribution system or is it integrated with the air handler? Check the ducts to and from the outside and make sure that they are open and clear of debris. Make sure that they are not blocked on the outside or clogged up on the inside. Make sure that during the original installation the openings into the HRV or ERV device were properly opened (these devices sometimes allow for a variety of installation configurations and a variety of openings through their housings). Make sure that the ducting from a bath fan runs all the way to the outside and that birds have not nested in the vent or that the uninsulated ducting in the attic is not full of water from condensation. Make sure that ducting running across joists is fully supported, not sagging down in an undulating wave across the attic floor. Check to see if all the duct connections or joints are tightly sealed with mastic or tape and are not hanging off.

How is it controlled? Is it designed to run on occupancy like a bathroom light switch, is it designed to turn on periodically throughout the day, or is it designed to run continuously? If it is coupled to the air handler there may be a control that monitors the operating time of the air handler to periodically open the outside vent and run the air handler. HRVs and ERVs are sometimes "hardwired" to run continuously with no control switch or they may have a power switch that is located near the fan unit. They may also have indoor air quality or humidity controls that will boost their airflow if the humidity or gases in the house increase.

Exhaust-only or supply-only systems may have associated "pin" timers that should be set up to operate the fan at particular times during the day.

After performing a visual inspection of the system and determining its desired operating configuration, flow measuring and system balancing can be done. Note that in measuring the flow through an exhaust-only or supply-only system, the measurements are being made on only one side of the system. These systems rely on leakage either in or out. Their effectiveness at accomplishing their purpose will be impacted by other devices operating in the house. Making flow measurements of an exhaust fan in a tight house with the clothes dryer running will reduce the flow through the bath fan because both devices will be competing for the same make-up air. To get a reasonable estimate of a particular device, other ventilation related devices should be defeated while the testing is taking place. That means shutting off the dryer, air handler, and combustion appliances that rely on house air for moving the air up the chimney. If you do shut off or subjugate such devices, be sure to liberate them by turning them back on before leaving the house being tested.

Flow Testing an Exhaust-Only System

Some sort of measuring device will be required to actually ascertain a numerical rate of flow through a fan or grille attached to a remotely mounted fan. Determining whether or not air is moving at all can be as simple as using a tissue, smoke pencil, or back of the hand.

If the measurement is made with a single point device like a mini-vane or hot wire anemometer, the reading will usually be in feet per minute (fpm), which can then be converted to cubic feet per minute (cfm) by multiplying by the area of the opening.

$$Q = VA$$

where

Q = the airflow in cubic feet per minute (cfm)

V = average velocity of the airflow in feet per minute (fpm)

A = cross sectional area of the duct in square feet (ft^2)

To determine the average velocity, divide the opening into equal areas and sample the velocity in each area. If the velocity profile is relatively flat, only a few points need to be sampled. A rapid traverse across the duct in two dimensions will define the uniformity of the airflow. If there are significant variations, use the mean velocity between the two extreme points.[4] The velocity profile is generally more uniform on suction openings.

The air will be moving into the grille or fan at different rates at different points across the face of the grille, so multiple measurements need to be taken, averaged, and then multiplied by the area to come up with a reasonable flow measurement. For a surface mounted grille, taking nine measurements will give a good approximation of the flow: the middle, the edge at four opposed compass points, and four points halfway between the middle and the four compass points.

Calculating the actual open area of the grille is not always easy. It can vary between 40 and 90 percent of their face area.[5]

Approximate free areas of different grille designs:

- "Eggcrate" (similar to the grilles in fluorescent light fixtures)—up to 90 percent open (see Figure 9.5)
- Pressed steel (metal grilles with straight vanes)—up to 75 percent open (see Figure 9.6)
- Fixed louvre (these can be metal or plastic with angled vanes; the plastic vanes are generally thicker and more restrictive)—30–70 percent open
- Double deflection (vanes pointing in both directions)—up to 60 percent open.

Although these are reasonable approximations, actual measurement of the particular grille in question would be required to get an accurate opening area to refine the accuracy of the measurement calculation. Calculate the face area of the grille (length times width or area of the circle) and multiply by the opening percentage (for example, pressed steel at 75 percent) to get the "clear" area. Multiply that times the average velocity flow reading to arrive at a cfm measurement through the fan.

If laboratory flow for the model of the fan can be checked against this tested reading, a level of reasonability can be achieved. For example, if the fan is rated to move 120 cfm and the reading is 150 cfm, the calculations need to be

4. A good discussion of measuring the airflow in ducts, pipes, hoods, and stacks can be found at http://www.omega.com/green/pdf/AIRFLOW_MEAS_REF.pdf.
5. http://www.vent-axia.com/knowledge/handbook/section2/grilles.asp.

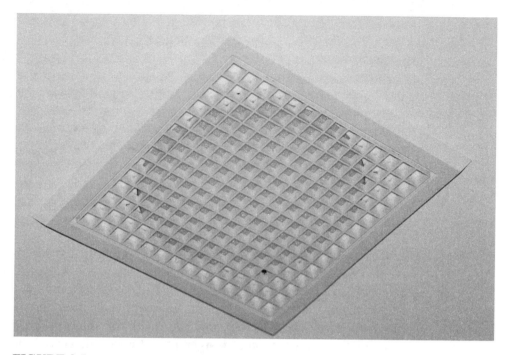

FIGURE 9.5   An "eggcrate" grille. (PHR)

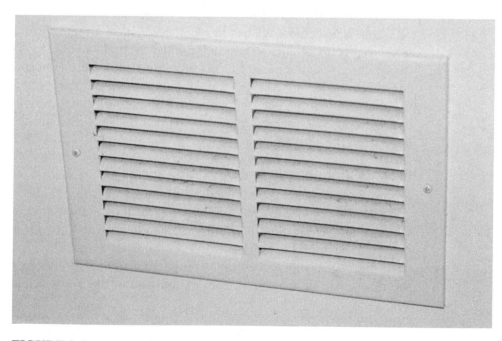

FIGURE 9.6   A pressed steel grille. (PHR)

Inputting the area in some digital mini-vane and hot-wire anemometers can provide a remarkably accurate measurement of flow without transverse measurements, averaging or carefully calculating the open area of the grille. Simply inputting the area of the ducting behind the fan or grille may be enough. In some tests, entering the actual open area of the grille which was approximately one half the area of the duct, provided flow measurements that were one half the actual flows as measured with a powered anemometer. The problem, of course, is knowing when to enter the open area and when to use the ducting dimensions. If a high degree of accuracy is required, taking several measurements with several devices will provide the best information.

rechecked. Installed system performance for most bath fans (because of the resistance of the ducting, etc.) is commonly about one half of their rated performance, so a 60 cfm measurement of a 120 cfm fan would not be out of line.

Devices like large vane anemometers that have capture hoods can provide a more accurate reading because the need for multiple measurements is removed, the capture hood is designed to allow all the air flowing through the fan to flow through the hood, and the conversion from velocity (fpm) to volume (cfm) can be automatic. Determining the actual open area of the grille, however, may still affect the accuracy of the calculation.

Using a powered duct-testing device will provide the most accurate flow measurement. The duct tester fan is connected to the exhaust fan opening. The flow through the duct tester fan is increased until there is no pressure difference between the air going into the fan under test and the room. At that point, the flow through the duct tester fan will equal the flow through the fan under test. This approach removes any impact that the flow-measuring device might have from the equation.[6]

Flow Testing Supply-Only Systems

Supply-only systems pressurize the building. They can be stand-alone devices delivering air to single or multiple points or they can be ducts from the outside that supply air directly into the air handler, using that system to circulate the air throughout the house.

If the system supplies air to the air handler through a duct from the outside, the flow could be measured from the outside of the house. Commonly, these

6. The Energy Conservatory's Duct Blaster® can be used for this process very effectively. Refer to Chapter 13.

systems don't have their own fans so the flow is initiated by turning on the air handler. Make sure that if there is a motorized damper in the inlet pipe that it is open, and measure the flow at the hood on the outside of the house. It may be difficult to get an accurate reading from the outside of the house if the inlet is near the top of the house and the siding does not allow a good seal around the edges of the flow hood due to ridges in the clapboard or shingles.

A second approach is to drill a hole in the ducting that leads from the outside and use a hot-wire or mini-vane anemometer to take a series of readings across the duct, averaged and multiplied by the area to come up with a cfm volume measurement. The hole should be drilled away from any sources of turbulence like elbows or dampers (4 feet if possible). At least five measurements should be made across the duct—against the wall, halfway between the wall and the middle, in the middle, halfway between the middle and the opposite wall, and at the opposite wall. Be sure that the hot-wire or mini-vane are as perpendicular to the flow as possible. New mini-vane anemometers, however, are fairly tolerant of yaw and pitch in facing the flow. Angle changes of up to 10 percent will result in less than 1 percent error, which is certainly acceptable in residential ventilation measurements.

A third approach would be to use a pitot tube and a digital or analog manometer (see Figure 9.7). A Pitot tube is an "L" shaped pipe that has a small hole at the end of the "L" for measuring total pressure and small holes on the sides of the "L" for measuring static pressure. At the top of the "L" there is a connection for reading the total pressure, and near the top of the "L" at the side there is a connection for reading the static pressure of the air moving through the duct.[7] The two connections are made to the two ports on the manometer and the result is a measurement of the "velocity pressure."

Some digital manometers like the Energy Conservatory's DG-700 can be connected directly to the pitot tube and will calculate the velocity (fpm) of airflow directly using the equation:

$$V = 1096.2 \times \left(\frac{V_p}{0.075} \right)^{0.50}$$

where

V = velocity of the airflow in feet per minute (fpm)

V_p = the velocity pressure in inches of water gauge

0.075 is in pounds per cubic foot (density of air at sea level)

---

7. Note that velocity pressure cannot be measured directly. It can be calculated from the equation: velocity pressure = total pressure − stack pressure.

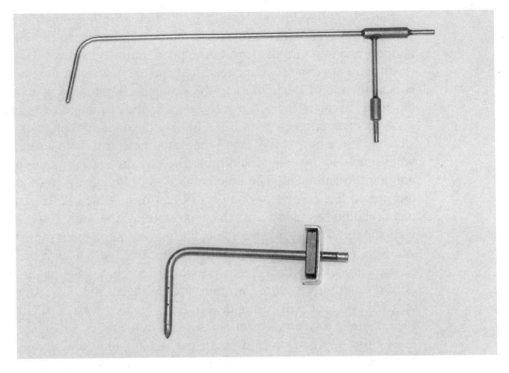

FIGURE 9.7   Pitot tubes. (PHR)

The resulting velocity measurement is then multiplied by the area of the duct to arrive at a flow measurement.

A fourth approach, and perhaps the most accurate way, is to drill a hole in the duct as you would for the hot-wire or pitot tube testing, and measure the "operating pressure" in the duct. This is similar to measuring the pressure in a room by pushing the hose under the door.

Once this pressure measurement has been made and recorded, the duct should be disconnected from the return plenum and connected to a calibrated fan or duct-testing device. Increase the flow through the duct tester until the same pressure with respect to the room pressure is achieved. At that point the flow through the calibrated fan will be the same as the flow through the duct when it is attached to the air handler. And don't forget to reconnect the duct to the air handler. You can get an estimated flow based on pressure at the FanCycler Website.[8]

8. The steps involved in this process are detailed at:
   http://www.fancycler.com/faq/default.htm#Outside%20air%20estimate.

Flow Testing Balanced Systems

Balanced systems must be tested for both the air coming in and the air going out. They rely on mechanical devices to move the air in both directions, so both mechanical devices must be doing the same work: they must be balanced! Commonly this happens in the same housing such as with a heat or energy recovery ventilator. These systems sometimes have just one fan motor with two "wheels" moving the air—one blowing in and one blowing out. Sometimes they have separate but similar fans. However, the total air delivered and the total air volume removed from the home depends not only on the mechanical devices pushing or pulling the air, but also on the ducting, the delivery system that guides the air to and from the rooms and to and from the outside of the house. Field studies have shown that over 50 percent of existing systems are in need of flow balancing.[9]

Symptoms of an unbalanced HRV/ERV could show up as:

- Frosting up of the HRV/ERV and/or the ducting;
- Supply air feels cool;
- Poor airflows;
- Humidity levels too low;
- Humidity levels too high.

The "garbage bag test" mentioned earlier can be used to get a good indication of the system balance.

1. Use a large garbage bag that is 48 inches long (1.2 m) and tape the opening to a wire coat hanger to keep it open.
2. Crush the bag and then hold it over the exhaust hood. The exhaust air will inflate the bag in about eight seconds. (If it takes less than that, set the control on the HRV/ERV to a lower flow setting and do the test again.)
3. Swing the bag through the air to inflate it, and then hold it over the supply or intake hood. The bag should deflate as quickly as it inflated.

If you find that the bag inflates much more quickly than it deflates you should try to determine the cause. Make sure all the duct connections are tight and that the filters are clean. It's amazing what you can find inside an HRV/ERV housing when the system has been running for a while. If those observations don't turn up any obvious problems, the system may need to be rebalanced using somewhat more sophisticated equipment.

9. HRAI.

Note that a "balanced system" does not just have to be a single package system like an HRV or ERV. A supply-only system like a controlled supply into an air handler can be combined with an exhaust-only system like a central exhaust fan to operate as a balanced mechanical ventilation system. It is important to match the flows on the two systems so they will work in harmony. It is also important to use a control strategy so the two systems will work together.

HRV and ERV manufacturers realize the importance of having their products work in a balanced manner and some of them provide flow taps on their systems or offer flow stations (a short length of duct and an air-velocity sensing grid). Along with the flow station, an analog (like a Magnehelic gauge) or digital manometer is needed to measure the pressure. The following procedures describe balancing a fully installed system and assumes that the house is complete, the HRV or ERV is fully installed, and any other appliances that move air into or out of the house (like a clothes dryer or range hood) are fully installed and operational.

Before you start balancing:

1. Fireplace dampers and windows and doors must be shut and tightly closed. Passive or mechanical inlet vents should be sealed.
2. Mechanical exhaust devices must be turned off.
3. Check the filters and make sure they are clean.
4. Make sure the exhaust and supply hoods are clear and open.
5. Make sure the HRV/ERV is not frosted up and the drain pan is not full of water.

Balancing Steps Using Flow Stations

Install the flow measuring stations. (Note that it is advisable to use two stations, one in each flow stream, because as the system is balanced, a number of readings will need to be made in both streams.) Air turbulence can have a significant effect on the measurements, so the flow measuring station should be located far enough away from turbulence producing elements (like elbows, dampers, or the fans themselves) to get accurate readings. As a rule of thumb, the flow measuring station should located a minimum of 12 inches (300 mm) away from dampers and 30 inches (750 mm) away from axial (propeller) fans or the outlets of blowers ("squirrel cage").

The flow measuring stations should be installed in the ducting referring to the arrows on the station for airflow direction. (Note that not all flow-measuring stations are direction sensitive.) The joints between the station and the ducting should be sealed.

Take the flow measurements:

1. Turn the HRV/ERV on high speed.

2. Measure the flow with the manometer. (Zero the gauge if required. A Magnehelic gauge must be vertical, not lying on its back. A slight tap on the gauge may be required before zeroing.)

3. Wait for the gauge to stabilize.

4. Convert the reading to airflow volume using the calibration label attached to the flow station.

5. If the reading has been made with an analog gauge, remove the gauge, disconnecting the hoses, and check that it returns to zero. If not, re-zero the gauge and retake the reading.

6. Repeat the procedure for the other air stream.

Balance the system: If the airflows are out of balance by 10 percent or more, a balancing damper in the air stream with the greater flow can be used to reduce the flow to bring it into balance. (If a balancing damper is used, the position of the damper should be locked in place once balance has been achieved.) If the higher flow rate is on the supply side, adjustable room grilles can be used to throttle down the volume. Note that the delivered airflow must still meet the design requirements once the flow has been reduced.

Balancing Steps Using a Pitot Tube

When it is difficult to remove sections of duct to install flow stations, measurements can be made using a pitot tube which only requires drilling a small (about ³⁄₁₆ inch) hole in the ducting. This procedure can be used effectively when the HRV/ERV is connected to the HVAC system or air handler that should also be running on high speed during the testing. This will provide the maximum pressure that the HRV/ERV will have to overcome.

1. Drill the measurement hole away from sources of turbulence (as with the flow measuring stations described previously);

2. Point the pitot tube into the airflow;

3. Connect the gauge. The connection at the top of the pitot tube (measuring the velocity pressure) should be connected to the high pressure side of the gauge. The connection at the side of the pitot tube (measuring the static

pressure) should be connected to the low pressure or reference side of the gauge. The gauge should be capable of reading from 0 to 0.25 in. wg (0 to 62.5 Pa).

4. Read the pressure in the supply side and then repeat the procedure on the exhaust side.

5. Determine which flow is the greatest.

6. Install and adjust a balancing damper in that stream to balance the flows. (Adjustable room grilles can sometimes substitute for a balancing damper.) Note that in heating climates, if an imbalance must be allowed, slightly higher exhaust than supply will increase the supply air temperature.

COMMISSIONING AND OWNER EDUCATION

Ventilation systems in homes can make a great deal of difference to the health and comfort of the occupants and the durability of the building itself, but only if they are running and operating effectively. There is absolutely no point in a beautifully designed and installed ventilation system if it shut off or defeated because the occupant of the house doesn't understand it.

Commissioning is the process that is used in commercial applications to make sure that installed systems work as they are supposed to work when the building is completed. The same process can be used for a residential application. With a new home there are two steps. One is making sure the systems work and the second is getting the owner to understand what they are doing, how they work, and how to maintain them. Good car dealers will carefully go over the systems in a new car when they transfer ownership. It is amazing how little new homeowners know about the systems in their homes.

As noted repeatedly in this book, the home should be reviewed as a system because everything should work effectively together and not interfere with each other. This is a book about residential ventilation systems, however. Commissioning the ventilation system essentially means making sure that it will work as it has been designed and installed.

Reviewing the system, as described previously, is the first step in the process. Simply making sure that the fans run and move air and the air moves from the inside to the outside of the house (and is not vented into the attic) is essential. The controls should work as they have been designed to work, boosting the flow, or activating the flow on motion, humidity, pressure, gases, or time. Timers should be programmed. If they have backup batteries, they should be fresh. (It is not uncommon during the construction process to activate and program timers and then shut the power off to the building for an extended time, longer than the backup batteries can handle.

For HRV/ERVs the connections to and from the outside should be checked to make sure they are not blocked. (It is not uncommon for the connections inside the exchanger box to be blocked.) The filters should be in place and clean. The core should be in place and clean. The controls should all be functional. (The installation manuals for these products have a great deal of helpful information on installation and testing. If there are questions, don't hesitate to contact the installer or the manufacturer. They want their systems to work well and they want the owners to be satisfied. They know how important word of mouth advertising is.)

A room-by-room check of air flowing into and out of the grilles can be done with a tissue or a smoke pencil or even a sensitive hand. Supply air should not be delivered at a rate that will cause uncomfortable drafts on the room occupants if it is particularly cold. If necessary, redirect the flows.

Check the condensation drain and the drain tubing.

After making sure the systems are running correctly, a walk-through with the homeowner will assist them in understanding what the systems are, how they work, and how to maintain them. There are so many systems in a house that the care and feeding of the ventilation system may play a surprisingly minor role in the homeowner's mind and memory, and they may not even know that something has gone wrong. It is the out-of-sight out-of-mind scenario. They want it to work and they don't want to think about it.

An exhaust-only system like a bathroom fan either runs or it doesn't. Pretty simple. But the fans are so quiet now that it is difficult to tell even in a quiet house whether they are working or not. The homeowner needs to understand how important it is that the exhaust-only fan is for the air quality in their home and periodically check with a tissue or some other means

Take, for example, setting a dehumidistat. Relative humidity is a difficult concept to begin with. Understanding that 75 percent RH is not the same as 75°F is not intuitive. Raising the set point on a dehumidistat will allow the humidity in the house to increase before the controlled device is activated. Raising the set point on a *humidistat* will allow the controlled device to run longer before it is deactivated, adding more humidity to the air. The ideal relative humidity levels in a house are between 35 and 55 percent RH. They can be a bit lower during the winter and higher in warmer weather.

that it is still running. Removing the grille to clean it is a reasonable way to accomplish this as well.

Sometimes a more dramatic demonstration of the relationship of the ventilation system to the air quality is required. A theatrical smoke generator in the garage is a dramatic way to show how the air flows from the garage into the house through the duct systems and is fully distributed.

A small humidity recording data logger (like the Hobo[10]) with the ventilation system off for a few days can provide a clear printout of the effect of air movement on humidity build-up. These devices are quite small and can be left in several places in the home, downloaded, and graphically printed out.

There are also companies like AirAdvice[11] or Aircuity[12] that will provide a multi-contaminant logger with a complete analysis of the pollutants in the home.

Sometimes it takes more than just knowing that the ventilation system is running to convince the occupant of its importance. And what about the next owner of the house? Leaving documentation that can be passed along that can also serve as a maintenance log will serve as an ongoing reminder.

PRODUCT TESTING AND LABORATORIES

It is simple enough to know whether a fan is moving any air. It is a different thing to know how much air it is moving and how much air it will move when it is on the job, pushing or pulling air through ducting. People have a sense of how much light a 40-watt bulb will produce because they can see it. How much air is 40 cubic feet per minute? As the importance of residential ventilation grows, as homes get tighter, the importance of knowing how the system will work when it comes out of the box and gets installed also grows. Without standardized, third party testing, a manufacturer can say anything they want about the product, and it may or may not move any air or exchange any heat and the homeowner wouldn't know because the moving air can't be seen.

To address this issue, the ventilation industry submits their products to a number of third party laboratories to test the products under standardized procedures for air movement, sound level, power consumption, and heat exchange efficiency.

10. www.onsetcomp.com.
11. www.airadvice.com.
12. www.aircuity.com.

HVI does not certify the sound levels of exterior mounted, in-line fans, or remote mounted heat/energy recovery ventilators. The sound level experienced by owners of these products will depend greatly on how they have been installed, the length and type of ducting, and the distance of the air mover from the interior grilles.

Residential ventilation product manufacturers addressed these issues by forming the Home Ventilating Institute (HVI) as a nonprofit trade association in 1955 (see Figure 9.8). HVI developed a series of standards and criteria for fan performance that could be repeated so that products could be directly compared to one another and so that consumers would have a scale to evaluate product A and product B. As of this writing, HVI has over 60 members and the Certified Product Directory is the place to find information on the performance of residential fans.

HVI uses the Texas Energy and Environmental Systems Laboratory (TEES) at Texas A&M University as the third party testing facility for residential fans and static vents and Bodycote[13] for testing heat and energy recovery ventilators. The product performance program has four elements: certification, verification, challenge and presentation of product performance ratings in the marketplace. Certification goes beyond just an initial test of the performance of the product. Every two years sample products are purchased in the marketplace by HVI and sent back to the laboratory for retesting to verify that the product is still performing as it was initially defined. And the performance of a certified product can be challenged, requiring a series of steps to confirm the performance or compel the manufacturer to make adjustments to bring it into compliance. To further assure the public of the quality of the HVI label, there are strict rules on the use of the logo on products and literature. Airflow testing is done under HVI Publication 916, sound testing under HVI Publication 915, Publication 920 defines the certification procedure, and HVI Publication 925 regulates labels and logos.[14]

Airflow testing at TEES varies depending on the product being tested, whether it's a ceiling mounted bath fan, remote mounted in-line fan, whole house comfort ventilator, through the wall fan, static vent, etc. All product

13. Bodycote Testing Group, www.bodycotetesting.com.
14. Visit the HVI Website for the complete testing procedures: www.hvi.org/resource-library/procedures.html.

FIGURE 9.8   A TEES test lab. (ESL)

testing is done in accordance with American National Standards Institute (ANSI) and Standards Council of Canada (SCC) consensus standards. It should be noted that HVI lists ratings in steps of 10 cfm. A fan that has a flow of 81 cfm and another fan that has a flow of 89 cfm are both rated at 80 cfm in the certified directory. From that standpoint ratings will always be on the conservative side.

The certification process requires having the fan safety tested (UL, ETL, CSA, Met Lab, etc.) and having the fan tested under the requisite airflow measurement setup. Sound testing is performed promptly after airflow testing without any modification to the product (see Figure 9.9). Sound testing is performed in a prescribed, diffuse, reverberation chamber using a reference sound source (RSS) and an extremely low background sound level. Four sound pressure measurements in 24, one-third-octave bands are conducted. "Sound power of the test unit is determined by mathematically comparing

The certified airflow rating is a great place to start to compare one fan to another, but the installed performance will differ from the certified rating point because of the resistance of the installation—the ducting and the termination fitting or hood or vent cap. Using a higher rating point like 0.25 in. wg will provide a better installed performance reference, although the actual installation resistance will probably exceed even that number.

sound pressure measurements in the chamber to sound pressure measurements of the RSS, using its sound power calibration data."[15]

The Certified Product Directory provides the following typical information:

- Model or series number: Manufacturer's product identification.
- Details: Specific product information like "30," "one piece model/crawl space model," or "VER. HS+" which means vertical high speed or "Hor. HS+" which means horizontal high speed for range hoods with either vertical or horizontal outlets that were tested at their high speed setting.
- Static pressure: For direct discharge fans the static pressure rating point is 0.03 in. wg because there is no ducting attached. For ducted fans the rating point is 0.1 in. wg with a second rating point at 0.25 in. wg being optional. Inline fans are rated at 0.20 in. wg with two additional rating points being optional.
- CFM: The airflow at the rating point. Alternative airflows can be listed at alternative resistances but the airflow at the rating point is what is listed on literature and promotion of the product.
- Sones: The sound level at the rating point in sones.
- Watts: The power consumption of the product is an optional number, but it is the power consumption in watts at the rating point.

HRV/ERV testing is performed at the Bodycote Laboratories in Ontario. These devices undergo long-term temperature testing because their purpose is to provide ventilation while saving heat. The air temperatures on the incoming air stream must be brought down below freezing so that moisture that collects in the system will freeze the way it would in an actual installation. (See Appendix G for a complete description of HRV/ERV testing.)

The result of all this testing is the entry in the HVI Certified Directory which provides an abundance of information on the performance of the product. The

15. HVI 915, Loudness Testing and Rating.

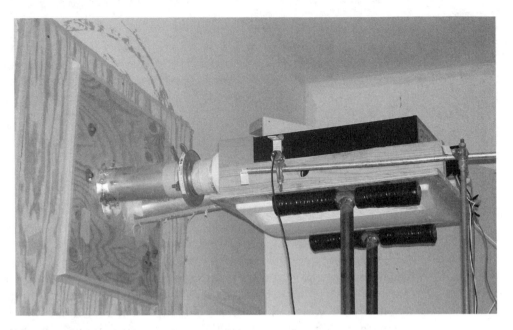

FIGURE 9.9 A room mounted ERV in a sound lab. (ESL)

"efficiency" of the system, its "performance," is the number that most evaluation systems are looking for. The HVI directory provides "Sensible Recovery Efficiency," "Apparent Sensible Effectiveness," and "Total Recovery Efficiency." It provides different numbers for heating at different airflows and other numbers for cooling.

These are the definitions in the HVI Directory:

> **"Apparent sensible effectiveness"** (ASEF) is "the measured temperature rise of the supply air stream divided by the difference between the supply temperature and exhaust temperature and multiplied by the ratio of mass flow rate of the supply divided by the minimum of the mass flow rate of the supply or exhaust streams. This value is useful principally to predict final delivered temperature at a given flow rate."[16]
>
> **"Sensible recovery efficiency"** (SRE) is "the sensible energy recovered minus the supply fan energy and preheat coil energy, divided by the sensible energy exhausted plus the exhaust fan energy. This calcu-

16. HVI Product Directory, "Heat Recovery Ventilators and Energy Recovery Ventilators," introductory page.

lation corrects for the effects of cross-leakage, purchased energy for fan and controls, as well as defrost systems. This value is used principally to predict and compare energy performance."

"Total recovery efficiency" (TRE) is "the total energy (enthalpy) recovered minus the supply fan energy and the preheat coil energy, divided by the total energy (enthalpy) exhausted plus the exhaust fan energy. This calculation corrects for the effects of cross-leakage and external purchased energy for fans and controls. It is used principally to predict and compare energy performance."

Since the motor or motors are in the same housing as the exchanger element the heat generated from the work they are doing is added to the air streams. The efficiency calculation should include that information so the SRE should be the number to use. The number listed in marketing information is usually the ASEF usually at the lowest operating speed and usually at 0°C or 32°F. Those specifications should be listed beside the efficiency percentages. It would be advisable to refer to the complete HVI directory for all the information prior to selecting an HRV or ERV. Talk to HVI or the product manufacturer for specific information if you have questions.

Chapter 10
SYSTEM TROUBLESHOOTING, SERVICE, AND MAINTENANCE

MY SYSTEM ISN'T MOVING ANY AIR

The first question to ask is: How do you know? What method have you used to measure it? The second question is: Did this happen suddenly or has it always been this way, not moving any air? The third question is: Is the air mover (fan or HRV/ERV) running at all? If it is a simple system—a bath fan on a light switch, for example, it will be much simpler to determine why it isn't working or why it isn't moving any air.

Sometimes the amount of air that is moving is difficult to measure because it is so low. The tissue or smoke methods described in Chapter 9 can give you a sense of subtle airflow. But the fact is that if the ventilation system is meant to move 90 cubic feet of air per minute and instead it is moving less than 20, there is definitely a problem that needs to be addressed. Knowing whether or not it is a new symptom or something that has been going on for a long time will give you a good place to start—whether a critter has recently moved into the ductwork or if the initial installation was defective.

Understanding the system is always the place to start. What sort of ventilation system is it? What are the component parts—for a fan: ducting, hood, or control, for an HRV: ducting, grilles, exterior inlet and outlet hoods, and controls. Understanding how the system is supposed to work will provide insight into why it is not working.

If there is no airflow or only occasional airflow, check to see if the air mover is running. (Occasional, subtle airflow could be caused by air pressures or wind blowing on the house.) But if there seems to be no steady airflow and no sound, chances are that the air mover isn't running. Fans are remarkably quiet these days so sometimes it is difficult to tell.

It could be that it is turned off. It could be that it is supposed to be off. Controls for ventilation systems are not always in obvious places, and they can be programmed to turn the system on and off periodically throughout the day. They are occasionally buried in closets or installed in an electrical box up near the ceiling, places where they have been hidden to prevent people from interfering with a designed ventilation strategy. Timers, for example, are generally solid mechanical devices, but they do fail eventually. Microprocessor based controls sometimes have battery backups to preserve their programming during a power failure. Although if the battery dies, they will be operating in their default mode—whatever the designer decided was the program that should take over when all else failed. Dehumidistats are misunderstood devices and can be set to work backwards—on humidity fall instead of humidity rise.

If it is a linear, wall-mounted humidity control (not an electronic humidity control that is built into the fan), rotate the setting counterclockwise to a very low setting like 40 percent RH (not all the way to off). If it is winter, it still may be too dry for the control to activate the fan. Try breathing into it and see if that gets it running.

If the control is a wall switch that controls both the fan and the light, and the light works, then the switch hasn't failed. A wire may have come loose or the fan may have failed. (Some new controls switch on both the light and the fan simultaneously but delay the off cycle of the fan.)

WARNING: Disconnect power to the system before attempting to service it. Electrical shocks can be damaging to your health!

The covers of bath fans can be removed for cleaning and that will also provide a means to visually determine if the wheel inside is spinning. Most bath fan covers are attached by "V" shaped spring clips located on either side of the grille. If there are no obvious screws, gently but firmly pull the grille down away from the fan. As you are pulling, the clips will become obvious. When the grille stops moving, it has reached the little loops on the ends of the spring clip wires. Compressing the ends of the "V" into each other will free the clips from the fan and allow you to remove the cover completely.

Is the fan spinning? If it isn't and whatever is controlling it says it should be, the fan is probably broken and needs to be replaced.

The covers to HRV/ERVs are designed to be removed to service the system, to clean the filters and the core particularly (see Figure 10.1). When you are up close and personal with an HRV/ERV, it is not a subtle device. The fans are designed to aggressively move a considerable volume of air through a considerable amount of ductwork, so if you remove the cover, the fans will stop because that's what they are designed to do for your safety. But before you remove the cover, if the fans are supposed to be running, you should be able to hear them. If you can, dig out the installation and service manual that came with the device. If you can't, most of the manufacturers have their product manuals on-line. And if you have trouble there, contact the manufacturer or the Home Ventilating Institute (HVI), and they will assist you in getting the documentation you need.

FIGURE 10.1   Servicing the core. (Fantech)

While you have the cover off the HRV/ERV, make sure the filters and core are clean and there isn't a pile of insects or other nasty things inside there that shouldn't be (see Figure 10.2). It isn't likely that you will have so much stuff in there that all the flow will be blocked, but crud will certainly reduce the airflow. And remember, what goes through this device is the good, clean, fresh air that you and your family are breathing—air that is supposed to keep you healthy.

If the air mover (fans, controls, HRV, or ERV) is working and there is still little or no airflow, the problem lies in the ducting. Does the ducting run all

Step back for a moment here and think about how you want your ventilation system to run. The original designer or installer may have had a completely different vision of how it should be working. Things might have changed. Teenagers can take longer showers than infants. You might be doing more cooking. So even if the system is running the way it was originally designed to run, it still might not be working the way you think it should and maybe that's why you're troubleshooting it. Perhaps humidity control does not suit your lifestyle no matter how it is set.

FIGURE 10.2 Servicing the core. (Venmar)

the way to the outside of the house and if so is the exterior vent blocked? Both the fan and the exterior vent are likely to have backdraft dampers either or both of which could be damaged and jammed shut.

If the hood isn't damaged or blocked, check the ducting. Someone may have stepped on it or dropped a trunk on it in the attic or it may be full of water or something else. If the ducting is crushed or damaged or if it is uninsulated ducting running through a cold attic, it should be replaced with smoother, better insulated, better supported ducting.

MY SYSTEM IS TOO NOISY/TOO QUIET

If the bath fan is too noisy, at least you know that's it running! If it's too quiet, maybe it isn't running or maybe it was designed to be really, really quiet. New fans that are rated at less than 1 sone can be so quiet that you might not know they are running. Noise is not the equivalent of airflow. Fans can be remarkably noisy and not be moving any air at all or remarkably quiet and be moving a lot of air.

If a fan is unacceptably noisy, it may just need to be replaced because that's the way it was made. Removing the grille (REMEMBER the warning about electrical power) should give you access to the product's model number. You can look up that model through the HVI certified product listing and find the sone rating or contact the manufacturer if it is no longer listed. If the fan is rated in excess of 1.5 sones, then it will be noisy no matter what other adjustments you might make.

If a fan is unacceptably quiet (and it is running), you might need to replace it with one with a higher sone rating. Sometimes a little fan noise in a bathroom or powder room is a good thing.

Fan noise can be increased by increasing the resistance of the airflow path. If the fan has to work harder, it is likely to make more noise, like blocking the nozzle on a vacuum. So if a fan that is designed to be quiet seems to be making more noise than it should, chances are that the ducting is twisted, kinked, or blocked in some way. Some fans have actually been designed to automatically increase their flow rates to compensate for duct resistance.

Noise level may vary as pressures on the house change, making the fan's job easier or harder, pushing on the exterior backdraft damper, for example. Giving the fan the easiest path from the inside to the outside of the house will provide the best performance and the quietest operation.

It should be noted that any noise is more noticeable in a tight house. If the ventilation system operates with automatic controls, it just emphasizes the importance of having the homeowner understand what it is, what it is for, and how it works. If they are sitting in the living room and the ventilation system suddenly shifts from an unnoticeable background level to high speed because some pollutant has been sensed, the homeowner may think that the system is defective and seek to defeat it altogether.

Exterior dampers can flap or bang in the changing pressures and breezes on the house. Removing the damper will certainly cure the problem, but the system is then relying on the primary damper in the bathroom fan (if there is one). Inserting a spring-loaded backdraft damper in the line will provide a better seal, but be sure that the fan can overcome the backpressure that the damper creates. A motorized damper in the line can cure the problem, but it will need to be controlled by the same control that controls the fan or by a pressure switch, sensing an increase in pressure in the duct and automatically activating the fan.

Fans can also produce motor hum that is transferred to the structure of the building. Motor hum is not part of the certification testing because it depends on how the installer installs the fan, which is not something the laboratory can verify. It is important that the installation steps described by the manufacturer be followed. Some bath fans, for example, require installing the housing, connecting the ducting, and making the electrical connections before inserting the working fan into the ceiling. That can make a difference.

If the sound level is critically important, isolating the fan from the structure is important. (Note that for the long-term satisfactory operation of the fan it is important that it be securely mounted.) There are a variety of sources for sound isolation hardware, with extremely descriptive Websites.

Remotely mounted exhaust or supply fans are very quiet because they are mounted remotely from the space they are ventilating! Sometimes, however, they are controlled with speed controls. It should be noted that not all fans should be controlled with solid-state speed controls. If the motor has not been tested for use with a solid-state speed control, it should tell you that either on the motor itself or in the installation manual. These controls generally provide only part of the AC electrical impulse to the motor, "chopping" the signal, turning the motor on and off rapidly. This process can be hard on the motor and also create motor hum. Sometimes the hum can be removed by adjusting the control to a slightly higher or lower speed. Sometimes these controls can also create interference with radio or television signals if they do not have adequate electronic filtration.

Because HRV/ERVs have so many different components, it is more complex to resolve sound problems. If the fans are running correctly and not making noises that would indicate that the fans themselves are damaged, then noise problems are likely to be transferred to the living space through the grilles. The problems could be with restricted flow paths through the ducting as described above for bath fans. The noise can be at the grille itself. If the velocity of the air being "thrown" out of the grille is high, it will make excessive noise. The grilles should have been matched to the system during the original installation so that the velocity of the air should not make excessive noise entering the room. If it does, chances are the system is unbalanced, that there is too much air being delivered to that particular room and not enough somewhere else. A balancing damper in the ducting can cure excessive airflow.

The motors in HRV/ERVs are certainly more powerful than the motors in typical bath fans so they can produce more motor hum. It is important to follow the manufacturer's installation instructions and not couple them directly to the structure of the building so that the structure itself becomes an amplifier. Some of these products come with hanging straps that will absorb motor noise or duct clamps with hanging straps that include rubber pads to absorb motor vibration.

MY SYSTEM IS TOO DRAFTY

This complaint may be the result of not understanding the system, it may be a poorly designed system, or it may be air coming back down the ducts from the outside.

Lack of understanding should be the least of these concerns. The system should not be designed or installed in such a way that the occupants are ever uncomfortable. Air movement can be sensed as a "draft" particularly if it is unconditioned outside air that is blowing on people. Ventilation requires air movement but it shouldn't be uncomfortable air movement.

In a residential installation, supply vents from positive pressure systems should be located above where people sit and walk and children play and crawl. The incoming air should have a chance to mix with the room air, and it shouldn't be moving at such a high velocity that people sense it as a draft. The vanes on these grilles should not be directed down into the living space. Some manufacturers have developed grilles that make use of the Coanda effect to deliver the air at the ceiling and compel the air to move across the ceiling, mixing with the room air.[1]

"Unbalanced" exhaust-only or supply-only ventilation systems rely on air leaking in or leaking out around the house. Those inlets and outlets should be so small and so distributed that air moving through them can be barely sensed, if at all. If the house is too leaky, drafts will occur from both "natural ventilation" and mechanical ventilation. Those unintentional leaks should be sealed from both a comfort and an energy use point of view.

1. http://www.fantech.net/accessories1.htm.

To prevent drafts from air flowing back into the house through a fan housing, most ceiling mounted bath fans include balanced backdraft dampers that allow the air to flow out but prevent the air from flowing back in. (Note that these systems rely on gravity to close. If the system has been designed for "ceiling installation," the damper is not likely to close if it is installed vertically in a wall.) Many exterior hoods include a second damper. These dampers sometimes are purposely designed to allow a small amount of leakage to flow through them even when they are closed. This small gap allows the positively pressurized air to start flowing. If the damper is completely tight all the way around its perimeter, it is possible that the damper won't open.

Spring loaded, butterfly dampers can be added to the ducting to provide a better seal. They will create more backpressure than the very light gravity damper that is usually included with the fan. These dampers can be tight enough that any other damper in the line can be removed, including both the exterior damper and the backdraft damper in the fan.

MY SYSTEM DOESN'T CLEAR THE MOISTURE OFF THE MIRROR

The mirror is cold. A shower creates steam in the air. The warm moist air hits the cold surface of the mirror and the moisture condenses and fogs the mirror. If the room temperature is high, the mirror temperature will be relatively warm. The air will be able to hold more moisture and the mirror is much less likely to fog. Raising the room temperature is probably the most effective way to keep the mirror from fogging.

If the airflow through the ventilation system is high enough relative to the amount of humidity being generated, it can haul the warm moist air out of the room before it has a chance to fog the mirror. (On average this is about 80 cfm of true airflow moving through the room.[2])

Localizing the intake vent in the proximity of the mirror for a remote mounted, in-line bath fan can drag the air directly over the mirror's surface to limit the amount of fogging that will occur (see Figure 10.3).

There are a variety of films that can be sprayed or wiped on the mirror to keep it from fogging. They even make a mirror for shaving in the shower. None of these are non-fogging because of airflow however.

2. Terry Brennan, Camroden Associates.

FIGURE 10.3 Window condensation. (Titon)

VENTILATION SYSTEM MAINTENANCE

Ventilation systems are designed to run for a long time with little conscious interaction with the occupant of the home. Because of that lack of interactivity, maintaining the system is not likely to occur until there are problems. The old English system of putting a shilling in the electric meter to get heat or hot water was probably the most direct feedback system. When the shower water started getting cold, you knew that your shilling had run out!

Component failures in HRV/ERVs are likely to begin when the system has been operating for four years. If low end bath fans are required to run continuously, their motors may fail in as little as one year.

System issues that are most likely to be problems are not likely to be noticed. For example, if one of the blowers in an HRV fails and the system is connected to the air handler, air will still be moving around in the house, but the effectiveness of the system will be severely compromised. Energy costs will increase because the house will rely on infiltration or exfiltration for system balance. Because of the problem of out-of-sight/out-of-mind, maintenance items can be overlooked.

Periodic maintenance like balancing the airflows in an HRV or ERV is not something that can be done by the homeowner without the proper equipment or training. It is extremely common for HRV/ERV systems to be out of balance either because they were never balanced in the first place or because the balance has not been maintained.

Homeowners commonly don't follow the instructions in the manuals that come with the products. Some routine maintenance issues are just as simple as changing the batteries in smoke detectors or the filter in the furnace air handler.

There is a disconnect between problems of indoor air quality (IAQ), moisture and the ventilation system. If the heat doesn't come on, the house gets cold. If moisture builds up, it is not obvious that the ventilation system isn't working.

The ventilation system is not a high priority in the minds of most homeowners when they buy an existing home, and the seller is not likely to explain the purpose or operating details of the system. The home's occupant needs to understand that it is good for their health to maintain the system.

If the homeowner is not familiar with the system and suddenly discovers that it is running when he/she thinks it shouldn't be (like when the windows are open and the weather is warm), they may shut the system off. When the cold weather comes around again, they may forget to turn it back on. Many homes have ventilation systems that have been defeated. The good news about these systems is that they generally can just be turned back on.

The operating condition of the equipment is important, but the understanding of the owner is equally important. If through lack of understanding the system has been modified (like adding an insect screen to the intake of an HRV) or the controls adjusted (like shutting off an exhaust-only fan that should be running continuously), the system will not function as it was designed to function. There is a lack of understanding about the purpose of the ventilation system. The fact that it is running all the time, blowing or forcing air out of the house seems like an obvious waste of money. That lack of understanding will lead to system "failure" as surely as a burned out fan motor.

Bath fan system maintenance

There aren't a lot of maintenance issues with a bath fan that is mounted in the ceiling of the bathroom. There's no filter to change and the fan motor is generally sealed and needs no oiling. If the fan works and moves air, clean the grille and leave the rest of it alone. (Unless the owner's manual for the particular fan in your ceiling says there are maintenance steps you should take.) If it is accessible, you should look at the exhaust or termination fitting on the outside of the house and make sure that there are no birds nesting in it or that the backdraft damper isn't jammed open or closed. You might want to check out the ducting to make sure that it hasn't been crushed or been disconnected from some fitting somewhere along the way.

Remote mounted or in-line fans can collect condensation inside their housings, particularly if they are mounted horizontally. Checking that means taking the system apart, which may not be a regular maintenance task. Other than that, checking the ducting and termination fittings as described previously is about all you need to do. If the fan has been in service for more than five years, it probably merits taking it apart and cleaning it. A filter can be added to the grille or ducting to prolong the life of the fan.

Exhaust-only systems should run for a long time without any maintenance or interference. Unfortunately with products like that, they tend to get ignored. They run quietly in the background until they are forgotten about altogether, and when they stop running, the effects may not be blatantly obvious. About the best possible maintenance that can be done is to make sure that the fan is still working.

Although this is not bath fan maintenance, supply-only systems should have filters that need to be changed as frequently as the local environmental conditions demand—maybe every year, maybe every three months. That filter is removing the particulate pollutants in the incoming air, and protecting the home's occupants.

HRV/ERV system maintenance

HRV/ERVs have more components and require more attention (see Table 10.1). They drag out the bad air, push in the fresh air, and run all that through the exchanger element. The outside air will contain dust, pollen, and other

The Canadian Mortgage and Housing Corporation (CMHC) has a clear simple description of basic HRV/ERV maintenance including a maintenance check list that you can copy and paste on the front of the unit to keep track and remind you of service issues. http://www.cmhc-schl.gc.ca/en/co/maho/gemare/gemare_004.cfm.

TABLE 10.1 Troubleshooting poor airflows. *(continued on next page)*

Symptom	Cause	Solution
Poor airflows	• ¼″ (6mm) mesh on the outside hoods is plugged • Filters plugged • Core obstructed • House grilles closed or blocked • Dampers closed (if installed) • Poor power at the site • Ductwork is restricting HRV/ERV • Improper speed control setting • HRV/ERV airflow unbalanced • Have contractor balance HRV/ERV	• Clean exterior hoods or vents • Remove and clean filter • Remove and clean core • Check and open grilles • Open and adjust dampers • Have electrician check supply voltage at the house • Check duct installation • Increase the speed of the HRV/ERV
Supply air feels cold	• Poor location of supply grilles, the airflows may be drafting on the occupants • Outdoor temperature extremely cold	• Locate the grilles on the walls or under the baseboards • Install ceiling mounted diffusers or grilles that do not spill the supply air directly on the occupants • Turn down the HRV/ERV supply speed • Add a small duct heater (1kw) to temper the supply air • Rearrange the furniture or open closed doors that are restricting the circulation of air in the home • If the supply air is ducted into the furnace return, the furnace blower may need to run continuously to distribute ventilation air comfortably
Dehumidistat is not operating	• Improper low voltage connection • External low voltage is shorted out by a staple or nail • Dehumidistat misadjusted or turned off	• Check that the correct terminals have been used • Check external wiring for a short • Adjust dehumidstat to the desired setting
Humidity levels are too high and condensation is appearing on the windows	• Dehumidistat is set too high • HRV/ERV is undersized to handle a hot tub, indoor pool, etc. • Lifestyle of the occupants • Moisture coming into the home from an unvented or unheated crawl space • Moisture is remaining in the bathroom and kitchen areas • Condensation seems to form in the spring and fall • HRV/ERV is set at too low a speed	• Set dehumidistat lower • Cover pools, hot tubs when they are not in use • Avoid hanging clothes to dry, storing wood, or venting clothes dryer inside (Wood for heating may have to be moved outside) • Vent crawl space and place a vapor barrier on the floor or seal up the crawl space and condition it • Ducts from the bathrooms should be sized to remove moist air effectively—Use bath fans after showers • On humid days in the changing seasons, some condensation may appear • Increase the speed of the HRV/ERV

TABLE 10.1   Troubleshooting poor airflows. *(continued from previous page)*

Symptom	Cause	Solution
Humidity levels are too low	• Dehumidistat set too low • Blower speed of HRV/ERV too high • Lifestyle of the occupants • HRV/ERV airflows may be out of balance	• Set dehumidistat higher • Decrease the HRV/ERV blower speed • House may have too many air leaks (This should be considered and addressed before adding any additional humidity to the space) • Have a contractor balance airflows
HRV/ERV and/or ducts are frosting	• HRV/ERV airflows are out of balance • Malfunction of the defrost system	• Note: minimal frost buildup is expected on cores before unit initiates the defrost cycle functions • Have HVAC contractor balance the airflows • If the unit has a self-test feature for the defrost cycle, use it to test the system
Condensation or ice buildup in insulated duct to the outside	• Incomplete vapor barrier around insulated duct • A hole or tear in the vapor barrier covering	• Tape and seal all joints • Tape any holes or tears in the vapor barrier covering • Ensure that the vapor barrier is completely sealed
Water in the bottom of the HRV/ERV	• Drain pans plugged • Improper connection of the drain lines • HRV/ERV is not level • Drain lines are obstructed • HRV/ERV core is not properly installed	• Clean out the drain pans • Check the connections • Check for kinks in the lines • Make sure water drains properly from the pans

Adapted from Aldes HRV Operation and Installation Manual.

particulates depending on the location and the local environment. If the system is not cleaned and maintained, it won't continue to work as it was designed to work.

Refer to the product manual and open up the housing for HRV/ERV air handler. That will provide you with access to the exchanger core and the filters. Clean out any detritus or dead bugs that might have collected in the housing. Clean and/or replace the filters. The grille in the kitchen should include a filter that should be maintained (cleaned or replaced) when the system filter is maintained. Additional filter boxes can be added to the lines to allow for the insertion of more effective filters. (Chapter 17 includes a section on furnace filters and MERV filter ratings. Those same ratings can be applied to ventilation systems.)

Pull out the core and carefully clean it. Check the fans to make sure both the supply and exhaust fans are still working.

Check on the condensate drain and pans. Make sure they are not blocked and the pans are clean. (Whatever is in those pans will eventually evaporate and be part of the airstream.) Make sure that there is a proper trap in the drain. If the HRV/ERV is connected to the air handler of the heating and cooling system, make sure that those connections are still solid. If they have been taped, consider resealing those joints with mastic. Some HRV/ERVs are designed to be coupled to the home's air handler without a "closed connections" (known as an "indirect connection") between the two, but deliver the outside air in the proximity of an inlet vent on the return side of the air handler's ducting. Often all of the incoming air is not "captured" by the opening in the return, particularly if the two systems are not control interlocked so that they both operate at the same time and the air handler fan is pulling in the air from the HRV/ERV at the same time as the HRV/ERV's blowers are running. This will result in over ventilating the basement and under ventilating the house. And if the air handler is located in an unconditioned basement, garage, or attic, it will depressurize that space, sucking in and distributing any other nearby pollutants. This sort of connection should be discouraged and repaired.

Check on the exterior hoods or termination fittings and make sure they are not blocked or don't have plants, leaves, or snow piling up in front of them. Make sure animals are not living in them and that their protective screens are still in place. Check the backdraft damper on the exhaust fitting. (There shouldn't be one on the supply fitting!) Make sure that there are no contamination sources like trash cans or rotting leaves under a deck near the intake. Homeowners will occasionally add insect screens to the intake vent out of a concern for insects being drawn into the system. These screens are very restrictive to airflow and will clog quickly. They are likely to be unnecessary because air and insects drawn into the system have to pass through the ducting, the system filter, the system core, through the blower, and out into the house. Hopefully the particulate filters in the system will be effective enough before something the size of an insect makes it all the way into the house.

Check the ducting:

- Poorly taped connections—poor tape, tape falling off, open connections;
- Poor sealing of the vapor barrier at the HRV/ERV and at the outside wall;
- Compressed ducting and insulation by hanging straps and unsupported runs;
- Missing insulation;
- Crushed flex where it has been squeezed into wall cavities;
- Poorly sized ducting and meandering duct runs.

Check the defrost system. The combination of cold outside air hitting warm humid air from the house will often result in frost build-up on the exchanger core. The defrost cycle in most systems activates when the temperature of the incoming air drops below 25°F (-5°C). Some systems have an internal motorized damper that temporarily blocks incoming fresh air, allowing only house air to circulate through and defrost the core. Other systems have an automatically timed supply fan shutdown. These systems are not easy to test unless the incoming air temperature is quite cold, but you can make sure that if there is an internal damper that it looks clear of dirt and debris. Do not try to move it manually as you are likely to strip the motor gears.

One final reminder: More than any other component in the house, ventilation systems in homes are there to keep the occupants and the building healthy. They must be maintained to accomplish that.

Chapter 11
COSTS OF VENTILATION

We have to accept the fact that residential ventilation is not free. Unless you live in what is affectionately termed a "paradise climate" where you can leave the windows open all the time and don't have to worry about excess humidity and you have no mechanical heating or cooling concerns, you're going to have to pay something to have air moving through your house. Just leaving the windows open will entail a cost for conditioning the air. So what's it worth to be able to breathe on a regular basis? (Of course, that's looking at the value, not the cost.)

There is a first cost for buying and installing the equipment. There is an energy cost for running the fan motors and there is an energy cost for conditioning the air moving through the house. The lifecycle cost includes all of these and adds the cost of replacing the system when it dies, removing the old equipment and taking it off to the dump or recycling facility. (It could even be extended in the other direction to include the cost of manufacture, delivery to the distribution facility, and finally to your house. It just depends on how far out you want to widen the circle.)

There are cost offsets that are virtually impossible to quantify. How much is saved for improved health for the occupants of the house with better ventilation? How much is saved for reducing the moisture effects on the structure of the house? It may be possible to quantify the savings in cooling load by reducing the temperature of the air in the attic with an attic fan (if it accomplishes that). A dehumidifier will reduce the humidity in the house, reducing the dehumidification load of the air conditioning, but it will also increase the temperature in the house, increasing the load of the air conditioning, and it will have an electrical cost.

Those are local costs. There are also global costs, carbon costs. In these terms, it is clear that the most efficient, most effective, longest lasting, simplest to maintain system with the least manufacturing and delivery cost will have the least overall cost to the planet.

The way prices for almost everything are changing, today's real numbers are not particularly meaningful here except in relative terms (particularly in terms of energy costs), so they are provided as an acceptable starting point.

CAPITAL COSTS

The mechanical ventilation systems in the house consist of a variety of components. At the heart is the system for the occupants, the primary ventilation system—the system that should be running all the time, exchanging the air throughout the house. This may be a stand-alone system or it might be integrated with the heating and cooling system. Then there are the spot ventilation systems—the bathroom and kitchen fans. There might (should) also be a radon or soil gas reduction fan system that runs all the time, extracting the soil gases from under the basement slab or around the foundation of the house. There might also be a garage ventilation fan controlling the build-up of carbon monoxide in the garage. There might be a whole-house comfort ventilator for warm weather cooling. And there might be an attic fan for removing the heat from the attic.

The associated costs for all of these systems and approaches are outlined in Table 11.1. The total first cost is the sum of the fan or device cost, the ducting, the hood or termination fitting on the outside of the house, the control (if it is separate), and an approximation of installation cost. (Note that these are just a few representative costs. Simpler installations would cost less; more complex would obviously cost more. There are a wide variety of all these systems with all sorts of installation variations.)

TABLE 11.1   Primary ventilation system first costs.

System	Installed Cost	System
Central exhaust-only	$535	Good quality exhaust fan with ducting and installation
Central exhaust-only with intakes	$885	Good quality exhaust fan with ducting, three intake vents, and installation
Central supply-only	$1,480	Central supply fan with distributed ducting to four rooms, control, and installation
Central supply to air-handler	$580	Ducting, control, damper, installation
HRV/ERV stand-only installation	$2,120[1]	Exchanger with fans, ducting, basic control, and installation
HRV/ERV attached to air handler	$1,520	Exchanger with fans, ducting, basic control, and installation

1. These numbers are for the primary ventilation system alone. An HRV/ERV exhausting from the bathrooms may eliminate the cost of the spot ventilation fans that would be necessary with a one-sided system.

The primary ventilation system for the occupants doesn't have to be expensive or complicated. If you consider the house to be a simple, fairly tight box of air, using a single point, exhaust-only ventilation system running continuously, it will change all the air in the box, diluting the polluted indoor air with fresher, outdoor air—dilution of the pollution. If the leaks into the house are generally evenly distributed and the house is simple enough and doors are left open, a central exhaust-only system can have a reasonably good impact on improving the air quality in the house.

The simplest approach is to use a centrally located bathroom-type fan. Very low cost, "builders' special," bath fans are not a good choice for this job. Some of these fans can cost under $10. They will be too noisy (in excess of 3 sones) to run very long. The occupants will find a way to turn them off. Indoor Environmental Quality (IEQ) includes the elimination of noise pollution, which becomes a more important factor in a tight house. A good quality bath fan will be more expensive, but the difference in up-front, installed cost will simply be the difference in the cost of the two fans. The installation details will be the same—same duct and same exterior hood or termination fitting (although a better fan may include better installation with larger duct and a lower resistance exterior hood). This fan could be located in a bathroom, serving as both the primary ventilation system and a spot ventilator for the bathroom. This works particularly well if a multi-speed controller controls the fan with one setting for the continuous background ventilation and a higher setting when the bathroom is occupied. (Some products do this speed shifting automatically with motion sensors or other occupancy indicators.)

An alternative to the ceiling mounted bath fan from a single bathroom is a remote mounted system that uses an in-line fan or a multi-port fan, drawing air from a number of points. If the system draws from a single point, its installation cost will be very similar to the ceiling mounted fan, with the cost of a grille added to the ducting, termination fitting, and control.

Some houses are tight enough to require the addition of make-up air or intake inlets (see Figure 11.1). For this analysis, these are passive devices located high on the walls or "slot vents" manufactured into the windows, allowing air to flow back into the house when it is under negative pressure from the primary ventilation system. They do not provide make-up air for any other purpose.

A remote mounted, in-line fan could also be used as a supply-only system, blowing air into bedrooms, for example. It is important to add a filter box to such a system and to locate the filter in an accessible location for maintenance. The first cost of this system would be similar to the remote system described above with the addition of the filter. However, this approach works

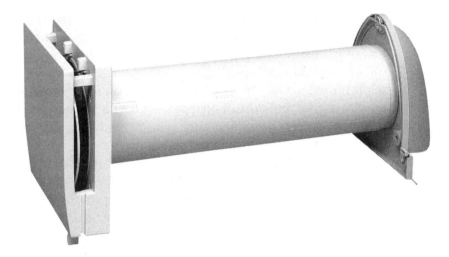

FIGURE 11.1   A through-wall intake vent. (Panasonic)

best if there are multiple distribution points, supplying air to several bed-rooms, which increases the cost of ducting, grilles, and installation.

A supply system added to the return side of the air handler is another inex-pensive approach. There are a number of supply side devices. Some systems are primarily a duct from the outside with a flow restrictor to limit the amount of outside air that is drawn in when the air handler fan is running. Others con-sist of a control system, ducting, and a motorized damper. The control moni-tors the operation of the air handler blower and logs run-time. It compares that run-time to the programmed-in desired ventilation run-time for the house. If the blower hasn't been running long enough to satisfy the ventilation require-ment, the control restarts the blower, drawing in outside air and circulating it throughout the house. The first cost on these systems is in the $90 to $300 range. Installation is relatively simple because it is localized to the area of the air handler and the controls are low voltage.

A stand-alone HRV or ERV is fully ducted, supplying fresh air to living spaces and extracting "used" air from polluting spaces like bathrooms (see Figure 11.2). Although these systems are more expensive, they can replace individual bathroom fans and their operating cost is reduced by their energy efficiency. Their first cost includes a complete ducting system that is less cost-ly than an HVAC ducting system because the ducts are smaller in size.

The cost of most of the ducting can be eliminated if the ERV is ducted to the air handler. Fresh, tempered air is delivered to the return side. Stale, house air can be removed from a point further upstream also on the return side, or

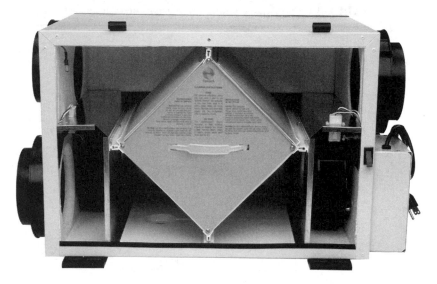

FIGURE 11.2   A heat recovery ventilator. (Fantech)

better yet, extracted from the bathrooms and kitchen. Spot ventilation fans may also need to be used to achieve adequate airflow from those spaces.

ELECTRICAL COSTS

Operating costs can be broken down into the electrical cost of operating the equipment and the cost of conditioning the air that is being removed and replaced (see Table 11.2). The systems that run virtually continuously, like the primary ventilation system and the radon/soil gas mitigation system, need to be electrically very efficient. Relying on a low cost bath fan that uses 90 watts of electricity would not be a good choice for a primary ventilation system running 24/7. In Hawaii with electricity costing 37 cents per kilowatt hour (kWh) in 2008, a 90-watt fan would have an electrical cost of $291.70 for the year. An 11.3-watt fan would cost $36.62 for the same year in the same place. At the average national electrical rate in 2008 of $0.1194 per kWh, a 90-watt fan would cost $94.13 and an 11.3-watt fan would cost $11.82. (The Energy Information Administration has good statistics on the cost of energy across the country for different sectors.[2])

2. http://www.eia.doe.gov/cneaf/electricity/epm/table5_6_a.html).

TABLE 11.2   Primary ventilation system electrical costs.

System	Wattage	Run Time (per day)	Annual Cost at $0.1194/hr
Central exhaust-only	11.3	24	$11.82
Central exhaust-only with intakes	11.3	24	$11.82
Central supply-only	26	24	$27.19
Central supply to air-handler	400 (air handler blower)	5.5[3] (average for ventilation air)	$95.88
HRV/ERV stand-alone installation	121	24	$126.56
HRV/ERV attached to air handler	121 + 400 (exchanger fan + air handler blower)	5[4]	$124.88

3. A primary factor in the operating cost of this system is the number of hours per day of run time or duty cycle.
4. The run time issue impacts this system as well.

Efficient bath fans operate at less than 15 watts. Inefficient bath fans consume over 90 watts. The electrical cost for the primary ventilation system is the power consumption of the fan in watts times the number of operating hours divided by 1000 (to convert to kWh) times the energy cost per kWh. For example a 30-watt fan running constantly ($24 \times 365 = 8760$) at $0.1194 per kWh:

$$(30 \times 8760)/1000 \times \$0.1194 = \$31.38 \text{ per year}$$

A 75 watt heat recovery ventilator would cost:

$$(75 \times 8760)/1000 \times \$0.1194 = \$78.45 \text{ per year}$$

If the furnace fan is used for the primary ventilation system and is run continuously, it may consume as much as 400 watts or more. This may be a good approach in terms of filtration and distribution, but if all the operating cost is attributed to ventilation:

$$(400 \times 8760)/1000 \times \$0.1194 = \$418.38$$

This is just the electrical cost! Of course a significant portion of the furnace blower's run time will be working to distribute conditioned air through-

A 100-watt light bulb uses 100 watts of power. If it runs for 10 hours, it uses 1000 watts, equivalent to 1-kilowatt hour represented as 1 kWh. Electricity is sold in kWh. In 2008 1 kWh in Hawaii cost $0.35. The same kWh costs $0.073 in Idaho with a national average of $0.1194. With the price of energy sashaying all over the place, it is difficult to predict an exact cost. Even if 1 kWh is relatively cheap, accumulating a lot of them can run the price up. "A billion here, a billion there, and pretty soon you're talking real money"– attributed to Everett Dirksen.

out the house, but if only 18 percent of the electrical cost is related to ventilation, that is $75 (at $0.1194/kWh).

The electrical cost for any of the other ventilation systems can be calculated in a similar way. A radon mitigation fan running continuously at 19 watts would cost $19.87 per year. A garage fan running continuously at 25 watts would cost $26 per year. A whole house comfort ventilator that uses 740 watts may only be running for eight hours per day over a three month or 90 day period or a total of 720 hours.

$$(740 \text{ watts} \times 720 \text{ hours})/1000 \times \$0.1194 = \$63.62 \text{ for the season}$$

All of these calculations are at the national average electrical cost of 2008. In some parts of the country, the electrical cost is going to be twice this amount. You will need to check your electric bill for the total delivered electrical cost to get an accurate number.

CONDITIONED AIR COSTS

When you blow conditioned air out of the house, it must be replaced with unconditioned, outside, fresh air. That cost is dependent on the location of the house, on the amount of air moved, and on the percentage of that air that

If the watts aren't obvious in the specifications, multiply 120 volts times the amps. For example, if the fan is rated at .75 amps: 120 × .75 = 90 watts. This is not a perfect conversion, but it is satisfactorily close.

would have moved naturally into and out of the house through infiltration and exfiltration (see Table 11.3). The cost of conditioning the air moving naturally into and out of the house is typically 20 to 30 percent of the heating load. Adding a mechanical ventilation system will increase that cost slightly, perhaps 5 or 6 percent. So in terms of heating load, for example, if the house costs $1000 per year to heat, 25 percent is from infiltration, or $250. Mechanical ventilation might add $60. A 75 percent efficient HRV would reduce the mechanical ventilation conditioning cost for heat to $15. Per Year! That's not a lot of money to breathe good air. The cost of mechanical ventilation is small compared to the cost of conditioning unpredictable infiltration in most homes.

There are other factors, like cooling cost, which would increase the conditioning cost, and reducing the run time, which would reduce the cost. If the house is tighter, the percentage of the conditioning cost related to ventilation will increase, but the dollar amount won't change a great deal.

Other ventilation systems besides the primary system will add to the conditioning costs but to a lesser degree. Bathroom spot ventilation systems generally operate for a very short period of time, often not much longer than when the bathroom is occupied. If an 80 cfm bath fan operates 30 minutes per day on average, that would be 876,000 cubic feet of air moved out of the house, at a constant 40 degree temperature difference that would be about 6 gallons of oil or 595 cubic feet of gas.

TABLE 11.3   Primary ventilation system conditioned air costs.

	Degree Days	Exhaust or Supply-only[5]	HRV (65% efficiency)
Boston Heating	5634	$150.93	$52.82
Boston Cooling	777	$69.46 ($0.181/kWh)	$24.31 ($0.181/kWh)
Washington Heating	4224	$127.39	$44.59
Washington Cooling	1075	$59.98 ($0.125/kWh)	$20.99 ($0.125/kWh)
Houston Heating	1396	$42.10	$14.74
Houston Cooling	2893	$178.69 ($0.125/kWh)	$62.54 ($0.125/kWh)
Phoenix Heating	1765	$53.23	$18.63
Phoenix Cooling	4355	$222.59 ($0.104/kWh)	$77.91 ($0.104/kWh)

5. All 65 cfm of exhaust or supply air-conditioned. No allowance for natural infiltration/exfiltration offset. Some percentage of mechanical ventilation air will take the place of air that would otherwise naturally infiltrate and exfiltrate from the house.

An exhaust-only primary ventilation system is a bit more complex to evaluate in terms of conditioned air cost. It goes back to the discussion on building pressures. An exhaust-only system will offset some of the natural infiltration that would have been moving in and out of the building even if the fan wasn't there, so not all of the conditioning cost can be attributed to the fan. "Under peak [weather] conditions, the standard house will have greatly increased driving forces and infiltration will be very high at the worst time. In tight houses with mechanical ventilation, infiltration is not driven by wind and cold but is dominated by the fan."[6] In many cases, about 50 percent of natural infiltration is offset by exhaust-only ventilation, giving it an average 50 percent heat recovery component, or a conditioning cost per the HRV example of $30 per year.

6. Kelley, "Ventilation System Performance in Energy Crafted Homes," 1995.

A range hood exhausts air for even less time per day (often because they are too noisy). So the cost of the conditioned air they displace is very small.

A radon or soil gas mitigation system draws its air from under the slab and from outside of the house. Some basement or crawl space air may be drawn into the system depending on how it has been designed, and in some cases, that will be conditioned air. Properly designed, the amount of conditioned air drawn into the system is small enough to ignore.

A garage spot ventilation system will have a similar impact on the cost of conditioned air. Garages are rarely conditioned spaces, so the air that is vented out will primarily be leaking into the garage from the outside. Some air may come from the house, but those gaps and leaks should be carefully sealed and minimized.

Whole house comfort ventilators purposely move outside air in and inside air out at a high volume. They decrease the cooling load from a temperature standpoint because the moving air makes the occupants more comfortable at a higher temperature, but they may increase the dehumidification load depending on the local climate and use conditions.

Attic exhaust fans should not be removing air from the house, but cycling outside air into the attic, through the fan, and back to the outside so they have little or no effect on conditioning loads unless the AC system is located in the attic.

MOTOR LIFE/REPLACEMENT COSTS

The motor is the key mechanical component in the fan. Aside from cosmetics, it is unlikely that the fan will be replaced until the motor burns out. If the fan is operating as the primary ventilation system, it is vital that it continue to run. It should be replaced as soon as it stops working; therefore the longevity of the fan motor has to figure into the operating cost of the product.

The fan motors on inexpensive fans were never expected to run for exceptionally long periods of time in homes.[7] Until people began to understand the need for mechanical ventilation, it was expected that bath fans would be used only about seven minutes per day on average. The motors in inexpensive fans may burn out in a year or two if they are allowed to run continuously.

It is difficult for fan and motor manufacturers to predict the life expectancy of their products (see Table 11.4). There are many variables involved. Much of it comes down to the type of bearings and the lubricants used. This is one of those areas where if you pay more, you get more. It costs more to make a better motor with better bearings and better lubricants.

There are a wide variety of motor types. Without taking the fan apart, it is often difficult to tell what sort of motor is spinning the blade. It is difficult to get a fan manufacturer to define the number of operating hours for their products. If all fan manufacturers defined the number of hours of motor life, we could easily compare one product with another, but they would all have to be tested to the same specifications. There are some general rules of thumb you can use to guess the longevity of the motor. Note that although programs often require or recommend fans that have been "rated for continuous operation" there is no official program to certify that feature.

TABLE 11.4   Typical motor hours of running life.

Motor Type	Hours of running life
C-frame Shaded pole	20,000 to 30,000
Permanent Split Capacitor (PSC)	20,000 to 30,000
Motorized impellers	50,000 to 70,000
Electronically Commutated Motors (ECM)	80,000 to 130,000

7. In point of fact, the inexpensive fans that have been installed in filling stations, restaurants and bars run so long and are so numerous exhaust fan manufacturers paid close attention to fan motor life.

After taking the grille off a low cost fan, if the part of the motor that spins is imbedded in the jaws of a "C," you can be pretty sure that is a "C" frame motor. The permanent split capacitor motors often have a capacitor like a small tin can strapped to their side or located somewhere else in the assembly. (Sometimes they look like a "fat domino" and are hidden in the wiring compartment.) Most of the in-line fans use motorized impellers. And if the fan uses an ECM motor, you can be sure that somewhere it will tell you because they are more expensive and very energy efficient.

There are 8760 hours in a year. In the worst case fans rated for 20,000 hours of life will hang in there for about two years and three months. Some fan motors are designed to run for 50,000 hours or more—well over five years. A fan that will be used infrequently as in a powder room doesn't need to have as long a life motor as one that will be run continuously. Furnace fans run about 33 percent of the time in their job of conditioning the air.

Table 11.5 provides a picture of expected fan lives based on the fan's use. Note that these are rough figures. The actual operating life of the fan will depend on many issues so there will be significant variations on both sides. Motor life is obviously not the only factor, especially if we're talking a hundred years!

TABLE 11.5   Ventilation product life expectancy.

Application	Hours Per Day	Expected Fan Life (by Fan Type)	Years of Operation
Primary ventilation system (high quality bath fan)	24	100,000	11.4
Primary ventilation system (ERV/HRV)	19.2 (80% run time)	70,000	10
Primary ventilation system (ECM motor) (for space conditioning only)	24	90,000	10
Furnace air handler	8	55,000	19
Furnace air handler (for continuous ventilation and space conditioning)	24	55,000	6.3
Radon/soil gas mitigation fan	24	64,000	7.3
Spot ventilator (powder room)	0.5	20,000	110
Range hood	.167 (10 minutes)	20,000	A very long time

Insulation materials have improved to the point where they last indefinitely, so motor life is governed by bearing life and bearing life is dependent on degradation primarily of the lubricant. Materials like steel degrade through chemical reactions like oxidation, which is accelerated by heat. Every 10°C approximately doubles the speed of the reaction. A motor designed so that the bearings are 10°C cooler will run about twice as long. Using these factors, a motor manufacturer can accelerate a life test on a motor by raising the temperature.

VENTILATION COST SUMMARY

The total cost of operating the ventilation systems in a home is the combined cost of the equipment, the electrical cost, the conditioned air cost, and the replacement cost for all the systems. Although there are definitely costs involved, the effects of good ventilation systems on the health of the occupants and the durability of the building make those costs worthwhile. The equipment may not be as glamorous as a high-end stove or a spa in the bathroom and cost avoidance is never obvious, but that certainly doesn't reduce the importance. And the costs are not that high.

It takes some work to figure out the cost of the system. Unfortunately there is no simple way of calculating the operating cost per year that could be tacked on the side of the box. Certified airflow from HVI will provide that component. Power consumption can be determined from the watts. Life expectancy of the motor can be guesstimated from Table 11.5 (or from the manufacturer if they will provide it). The run-time can be estimated from the application. The conditioned air cost depends on the application, the airflow, the local conditions, the heat recovery efficiency, and the run-time.

You can go through these calculations, but the bottom line is that ventilation is not a major cost component in maintaining a healthy home. It is difficult to accept the idea that if a fan is running all the time, that it isn't wasting energy, but there are many less essential components running constantly in the house that we are generally not even aware of. Clocks, power strips, transformers, stereo equipment, answering machines, computers, all use power all the time. Background power consumption will rarely drop below 300 watts even without a ventilation system running. That is $(300 \times 8760)/1000 \times \$0.1194 = \$313$ per year. And if you add air conditioning and heating, water heaters, stoves, and refrigerators and freezers to that, you will get your total electrical cost. We just accept those costs as part of life. Reducing the cost of energy for the house by building it tightly allows the cost of operating a primary ventilation system at less than a dollar a day to be acceptable.

Chapter 12
VENTILATION CODE REQUIREMENTS

The most important building code issues are those that apply locally to the building that you are working on. There is no national, universal building code that applies to every house in every neighborhood. States and local jurisdictions have adopted and adapted various national and international building, mechanical, and energy codes.

The International Code Council (ICC)[1] is "a membership association dedicated to building safety and fire prevention, develops the codes used to construct residential and commercial buildings, including homes and schools." On its Website is a map identified as "I-Code Adoptions" that allows state by state identification of what "I" codes have been adopted. For example, as of this writing Massachusetts has adopted the 2003 International Building Code (IBC), the International Mechanical Code (IMC) and the 2003 International Residential Code (IRC). Massachusetts adapted the 2003 IBC as the basis for 780 CMR, the State Building Code.

Another example is the town of Rockford, Illinois has adopted the 2003 IBC and the 2003 IRC and added local amendments to both of them. The ICC bookstore sells the code books for states like North Carolina, California, Michigan, Virginia, Florida, Ohio, and Arkansas. Although most of the local codes are based on the ICC codes, it is important to check with local code authorities for the most pertinent codes, those that apply to the building you are working on. Some states like California, Florida, Washington, Vermont, and Minnesota have specific ventilation codes or procedures.

On the home page of the International Code Council there is a link to a map of the U.S. that identifies which states have adopted which codes. On the Department of Energy's Website there is also a map that indicates the status of state energy codes.[2] (Click on your state to determine which codes are applicable.)

(NOTE: BOCA stands for Building Officials and Code Administrators International)

1. http://www.iccsafe.org.
2. http://www.energycodes.gov.

IMPORTANT NOTE: The following are excerpts from code language that apply to residential ventilation applications. They are not meant to be the final and comprehensive word on the codes that apply in your application at this particular moment. These excerpts provide a good starting point, but the actual codes and local code officials should be consulted.[3]

2009 INTERNATIONAL MECHANICAL CODE

IMC § 303.3 Prohibited locations

Fuel-fired appliances shall not be located in, or obtain combustion air from, any of the following rooms or spaces:

1. Sleeping rooms.
2. Bathrooms.
3. Toilet rooms.
4. Storage closets.
5. Surgical rooms.

Exception: This section shall not apply to the following appliances:

1. *Direct-vent appliances that obtain all combustion air directly from the outdoors.*
2. *Solid fuel-fired appliances, provided that the room is not a confined space and the building is not of unusually tight construction.*
3. *Appliances installed in a dedicated enclosure in which all combustion air is taken directly from the outdoors, in accordance with Chapter 7. Access to such enclosure shall be through a solid door, weather-stripped in accordance with the exterior door air leakage requirements of the International Energy Conservation Code and equipped with an approved self-closing device.*

IMC § 401.1 Scope

This chapter shall govern the ventilation of spaces within a building intended to be occupied. Mechanical exhaust systems, including exhaust systems serving clothes dryers and cooking appliances; hazardous exhaust systems; dust,

---

3. The ICC has permitted the inclusion of these code citations verbatim. The author would like to express his sincere gratitude. Please visit www.iccsafe.org for copies of the codes from their bookstore.

stock and refuse conveyor systems; subslab soil exhaust systems; smoke control systems; energy recovery ventilation systems, and other systems specified in Section 502 shall comply with Chapter 5.

IMC § 401.2 Ventilation required

Every occupied space shall be ventilated by natural means in accordance with Section 402 or by mechanical means in accordance with Section 403.

401.3 When required

Ventilation shall be provided during the periods that the room or space is occupied.

IMC § 401.4 Opening location. Air intake openings shall comply with all of the following:

1. Intake openings shall be located a minimum of 10 feet (3048 mm) from lot lines or buildings on the same lot. Where openings front on a street or public way, the distance shall be measured to the centerline of the street or public way.

2. Mechanical or gravity outdoor air intake openings shall be located not less than 10 feet (3048 mm) horizontally from any hazardous or noxious contaminant source, such as vents, streets, alleys, parking lots, and loading docks, except as specified in Item 3 or Section 501.2.1.

3. Intake openings shall be located not less than 3 feet (914 mm) below contaminant sources where such sources are located within 10 feet (3048 mm) of the opening.

4. Intake openings on structures in flood hazard areas shall be at or above the design flood level.

IMC § 401.5 Intake opening protection

Air intake openings that terminate outdoors shall be protected with corrosion-resistant screens, louvers, or grilles. Openings in louvers, grilles, and screens shall be sized in accordance with Table 401.5, and shall be protected against

TABLE 401.5 Opening sizes in louvers, grilles, and screens protecting outdoor exhaust and air intake openings.

Outdoor Opening Type	Minimum and Maximum Opening Sizes In Louvers, Grilles, and Screens Measured In Any Direction
Intake openings in residential occupancies	Not < ¼ inch and not > ½ inch
Intake openings in other than residential occupancies	> ¼ inch and not > 1 inch

For SI: 1 inch = 25.4 mm

local weather conditions. Outdoor air intake openings located in exterior walls shall meet the provisions for exterior wall opening protections in accordance with the International Building Code.

IMC § 401.6 Contaminant sources

Stationary local sources producing air-borne particulates, heat, odors, fumes, spray, vapors, smoke, or gases in such quantities as to be irritating or injurious to health shall be provided with an exhaust system in accordance with Chapter 5 or a means of collection and removal of the contaminants. Such exhaust shall discharge directly to an approved location at the exterior of the building.

IMC § [B]402.1 Natural ventilation

Natural ventilation of an occupied space shall be through windows, doors, louvers or other openings to the outdoors. The operating mechanism for such openings shall be provided with ready access so that the openings are readily controllable by the building occupants.

IMC § 403 MECHANICAL VENTILATION

IMC § 403.1 Ventilation system

Mechanical ventilation shall be provided by a method of supply air and return or exhaust air. The amount of supply air shall be approximately equal to the amount of return and exhaust air. The system shall not be prohibited from producing negative or positive pressure. The system to convey ventilation air shall be designed and installed in accordance with Chapter 6.

IMC § 403.2 Outdoor air required

The minimum outdoor airflow rate shall be determined in accordance with Section 403.3. Ventilation supply systems shall be designed to deliver the required rate of outdoor airflow to the breathing zone within each occupiable space.

Exception: Where the registered design professional demonstrates that an engineered ventilation system design will prevent the maximum concentration of contaminants from exceeding that obtainable by the rate of outdoor air ventilation determined in accordance with Section 403.3, the minimum required rate of outdoor air shall be reduced in accordance with such engineered system design.

IMC § 403.2.1 Recirculation of air

The outdoor air required by Section 403.3 shall not be recirculated. Air in excess of that required by Section 403.3 shall not be prohibited from being recirculated as a component of supply air to building spaces, except that:

1. Ventilation air shall not be recirculated from one dwelling to another or to dissimilar occupancies.

2. Supply air to a swimming pool and associated deck areas shall not be recirculated unless such air is dehumidified to maintain the relative humidity of the area at 60 percent or less. Air from this area shall not be recirculated to other spaces where 10 percent or more of the resulting supply airstream consists of air recirculated from these spaces.

3. Where mechanical exhaust is required by Note b in Table 403.3, recirculation of air from such spaces shall be prohibited. All air supplied to such spaces shall be exhausted, including any air in excess of that required by Table 403.3.

4. Where mechanical exhaust is required by Note g in Table 403.3, mechanical exhaust is required and recirculation is prohibited where 10 percent or more of the resulting supply airstream consists of air recirculated from these spaces.

IMC § 403.2.2 Transfer air

Except where recirculation from such spaces is prohibited by Table 403.3, air transferred from occupiable spaces is not prohibited from serving as make-up air for required exhaust systems in such spaces as kitchens, baths, toilet rooms, elevators, and smoking lounges. The amount of transfer air and exhaust air shall be sufficient to provide the flow rates as specified in Section 403.3. The required outdoor airflow rates specified in Table 403.3 shall be introduced directly into such spaces or into the occupied spaces from which air is transferred or a combination of both.

IMC § 403.3 Outdoor airflow rate

Ventilation systems shall be designed to have the capacity to supply the minimum outdoor airflow rate determined in accordance with this section. The occupant load utilized for design of the ventilation system shall not be less than the number determined from the estimated maximum occupant load rate indicated in Table 403.3. Ventilation rates for occupancies not represented in Table 403.3 shall be those for a listed occupancy classification that is most similar in terms of occupant density, activities, and building construction; or shall be determined by an approved engineering analysis. The ventilation system shall be designed to supply the required rate of ventilation air continuously during the period the building is occupied, except as otherwise stated in other provisions of the code.

Exception: The occupant load is not required to be determined, based on the estimated maximum occupant load rate indicated in Table 403.3 where approved statistical data document the accuracy of an alternate anticipated occupant density.

TABLE 403.3 Minimum ventilation rates (private dwellings, single and multiple).

Occupancy Classification	People Outdoor Airflow Rate In Breathing Zone cfm/Person	Default Occupancy Density #/1000 ft^{2a}	Exhaust Airflow Rate cfm/ft^{2a}
Garages, common for multiple units[b]			0.75
Garages, separate for each dwelling[b]			100 cfm per car
Kitchens[b]			25/100[f]
Living areas[c]	0.35 ACH but not less than 15 cfm/person	Based upon the number of bedrooms. First bedroom, 2; each additional bedroom, 1	
Toilet rooms and bathrooms[g]			20/50[f]

For SI: 1 cubic foot per minute = 0.0004719 m^3/s, 1 ton = 908 kg,
1 cubic foot per minute per square foot = 0.00508 m^3/(s • m^2),
°C = [(°F) -32]/1.8, 1 square foot = 0.0929 m^2.

Footnotes:
a. Based upon net floor area.
b. Mechanical exhaust required and the recirculation of air from such spaces as permitted by Section 403.2.1 is prohibited (see Section 403.2.1, Item 3).
c. Spaces unheated or maintained below 50°F are not covered by these requirements unless the occupancy is continuous.
f. Rates are per room unless otherwise indicated. The higher rate shall be provided where the exhaust system is designed to operate intermittently. The lower rate shall be permitted where the exhaust system is designed to operate continuously during normal hours of use.
g. Transfer air permitted in accordance with Section 403.2.2.

IMC § 403.4 Exhaust ventilation

Exhaust airflow rate shall be provided in accordance with the requirements in Table 403.3. Exhaust make-up air shall be permitted to be any combination of outdoor air, recirculated air and transfer air, except as limited in accordance with Section 403.2.

IMC § 403.7 Balancing

The ventilation air distribution system shall be provided with means to adjust the system to achieve at least the minimum ventilation airflow rate as required by Sections 403.3 and 403.4. Ventilation systems shall be balanced by an approved method. Such balancing shall verify that the ventilation system is capable of supplying and exhausting the airflow rates required by Sections 403.3 and 403.4.

IMC § 404 ENCLOSED PARKING GARAGES
IMC § 404.1 Enclosed parking garages
Mechanical ventilation systems for enclosed parking garages shall be permitted to operate intermittently where the system is arranged to operate automatically upon detection of vehicle operation or the presence of occupants by approved automatic detection devices.
IMC § 404.2 Minimum ventilation
Automatic operation of the system shall not reduce the ventilation rate below 0.05 cfm per square foot (0.00025 m^3/s • m^2) of the floor area and the system shall be capable of producing a ventilation airflow rate of .75 cfm per square foot (0.0076m^3/s • m^2) of floor area.

IMC § 405 SYSTEMS CONTROL
IMC § 405.1 General
Mechanical ventilation systems shall be provided with manual or automatic controls that will operate such systems whenever the spaces are occupied. Air-conditioning systems that supply required ventilation air shall be provided with controls designed to automatically maintain the required outdoor air supply rate during occupancy.

IMC § 406 VENTILATION OF UNINHABITED SPACES
IMC § 406.1 General
Uninhabited spaces, such as crawl spaces and attics, shall be provided with natural ventilation openings as required by the International Building Code or shall be provided with a mechanical exhaust and supply air system. The mechanical exhaust rate shall be not less than 0.02 cfm per square foot (0.00001 m^3/s • m^2) of horizontal area and shall be automatically controlled to operate when the relative humidity in the space served exceeds 60 percent.
IMC § 501.2 Exhaust discharge
The air removed by every mechanical exhaust system shall be discharged outdoors at a point where it will not cause a nuisance and not less than the distances specified in Section 501.2.1. The air shall be discharged to a location

Author's note: In the 2006 version of the code, the following sentence was included, which seems like a sensible approach despite the fact that it is no longer in the code. *For a mechanically vented crawl space design, there must be a continuously sealed vapor retarding ground cover, have no fixed ventilation openings to the outside, and include a continuously operating exhaust fan.*

from which it cannot again be readily drawn in by a ventilating system. Air shall not be exhausted into an attic or crawl space.

Exceptions:

1. *Whole-house ventilation-type attic fans shall be permitted to discharge into the attic space of dwelling units having private attics.*

2. *Commercial cooking recirculating systems.*

501.2.1 Location of exhaust outlets

The termination point of exhaust outlets and ducts discharging to the outdoors shall be located with the following minimum distances:

1. For ducts conveying explosive or flammable vapors, fumes or dusts: 30 feet (9144 mm) from property lines; 10 feet (3048 mm) from operable openings into buildings; 6 feet (1829 mm) from exterior walls and roofs; 30 feet (9144 mm) from combustible walls and operable openings into buildings which are in the direction of the exhaust discharge; 10 feet (3048 mm) above adjoining grade.

2. For other product-conveying outlets: 10 feet (3048 mm) from the property lines; 3 feet (914 mm) from exterior walls and roofs; 10 feet (3048 mm) from operable openings into buildings; 10 feet (3048 mm) above adjoining grade.

3. For all environmental air exhaust: 3 feet (914 mm) from property lines; 3 feet (914 mm) from operable openings into buildings for all occupancies other than Group U; and 10 feet (3048 mm) from mechanical air intakes. Such exhaust shall not be considered hazardous or noxious.

4. Exhaust outlets serving structures in flood hazard areas shall be installed at or above the design flood level.

5. For specific systems:
   - 5.1 Clothes dryer exhaust, Section 504.4.
   - 5.2 Kitchen hoods and other kitchen exhaust equipment, Section 506.3.12, 506.4 and 506.5
   - 5.3 Dust stock and refuse conveying systems, Section 511.
   - 5.4 Subslab soil exhaust systems, see Section 512.4.

IMC § 501.2.1.1. Exhaust discharge. Exhaust air shall not be directed onto walkways.

IMC § 501.2.2 Exhaust opening protection. Exhaust openings that terminate outdoors shall be protected with corrosion-resistant screens, louvers, or grilles. Openings in screens, louvers, and grilles shall be sized not less that ¼ inch (6 mm) and not larger than ½ inch (13 mm). Openings shall be protected against local weather conditions. Outdoor openings located in exterior walls shall meet the provisions for exterior wall opening protectives in accordance with the International Building Code.

IMC § 501.3 Pressure equalization

Mechanical exhaust systems shall be sized to remove the quantity of air required by this chapter to be exhausted. The system shall operate when air is required to be exhausted. Where mechanical exhaust is required in a room or space in other than occupancies in R-3 and dwelling units in R-2, such space shall be maintained with a neutral or negative pressure. If a greater quantity of air is supplied by a mechanical ventilating supply system than is removed by a mechanical exhaust for a room, adequate means shall be provided for the natural or mechanical exhaust of the excess air supplied. If only a mechanical exhaust system is installed for a room or if a greater quantity of air is removed by a mechanical exhaust system than is supplied by a mechanical ventilating supply system for a room, adequate make-up air consisting of supply air, transfer air, or outdoor air shall be provided to satisfy the deficiency. The calculated building infiltration rate shall not be utilized to satisfy the requirements of this section.

IMC § 504 CLOTHES DRYER EXHAUST
IMC § 504.1 Installation

Clothes dryers shall be exhausted in accordance with the manufacturer's instructions. Dryer exhaust systems shall be independent of all other systems and shall convey the moisture and any products of combustion to the outside of the building.

Exception: This section shall not apply to listed and labeled condensing (ductless) clothes dryers.

IMC § 504.2 Exhaust penetrations

Where a clothes dryer exhaust duct penetrates a wall or ceiling membrane, the annular space shall be sealed with noncombustible material, approved fire caulking, or a noncombustible dryer exhaust duct wall receptacle. Ducts that exhaust clothes dryers shall not penetrate or be located within any fireblocking, draftstopping or any wall, floor/ceiling, or other assembly required by the International Building Code to be fire-resistance rated, unless such duct is constructed of galvanized steel or aluminum of the thickness specified in Section 603.4 and the fire-resistance rating is maintained in accordance with the International Building Code. Fire dampers, combination fire/smoke dampers, and any similar devices that will obstruct the exhaust flow, shall be prohibited in clothes dryer exhaust ducts.

IMC § 504.4 Exhaust installation

Dryer exhaust ducts for clothes dryers shall terminate on the outside of the building and shall be equipped with a backdraft damper. Screens shall not be installed at the duct termination. Ducts shall not be connected or installed with sheet metal screws or other fasteners that will obstruct the exhaust flow. Clothes dryer exhaust ducts shall not be connected to a vent connector, vent, or chimney. Clothes dryer exhaust ducts shall not extend into or through ducts or plenums.

IMC § 504.5 Make-up air

Installations exhausting more than 200 cfm (0.09 m³/s) shall be provided with make-up air. Where a closet is designed for the installation of a clothes dryer, an opening having an area of not less than 100 square inches (0.0645 m²) shall be provided in the closet enclosure or make-up air shall be provided by other approved means.

IMC § 504.6 Domestic clothes dryer ducts

Exhaust ducts for domestic clothes dryers shall conform to the requirements of Sections 504.6.1 through 504.6.7.

IMC § 504.6.1 Material and size. Exhaust ducts shall have a smooth interior finish and shall be constructed of metal a minimum 0.016 inch (0.4 mm) thick. The exhaust duct size shall be 4 inches (102 mm) nominal in diameter.

IMC § 504.6.2 Duct installation. Exhaust ducts shall be supported at 4-foot (1219 mm) intervals and secured in place. The insert end of the duct shall extend into the adjoining duct or filling in the direction of airflow. Ducts shall not be joined with screws or similar fasteners that protrude into the inside of the duct.

IMC § 504.6.3 Transition ducts. Transition ducts used to connect the dryer to the exhaust duct system shall be a single length that is listed and labeled in accordance with UL 2158A. Transition ducts shall be a maximum of 8 feet (2438 mm) in length and shall not be concealed within construction.

IMC § 504.6.4. Duct length. The maximum allowable exhaust duct length shall be determined by one of the methods specified in Section 504.6.4.1 or 504.6.4.2.

IMC § 504.6.4.1. Specified length. The maximum length of the exhaust duct shall be 35 feet[4] (10,668 mm) from the connection to the transition duct from the dryer to the outlet terminal. Where fittings are used, the maximum length of the exhaust duct shall be reduced in accordance with Table 504.6.4.1.

---

4. The International Residential Code specifies the maximum length allowed to be 25 feet (7620 mm) in M1502.4.4.1.

TABLE 504.6.4.1   Dryer exhaust duct fitting equivalent length.

Dryer Exhaust Duct Fitting Type	Equivalent Length
4″ radius mitered 45-degree elbow	2 feet 6 inches
4″ radius mitered 90-degree elbow	5 feet
6″ radius smooth 45-degree elbow	1 foot
6″ radius smooth 90-degree elbow	1 foot 9 inches
8″ radius smooth 45-degree elbow	1 foot
8″ radius smooth 90-degree elbow	1 foot 7 inches
10″ radius smooth 45-degree elbow	9 inches
10″ radius smooth 90-degree elbow	1 foot 6 inches

For SI: 1 inch = 25.4 mm, 1 foot = 304.8 mm, 1 degree = 0.0175 rad.

IMC § 504.6.4.2 Manufacturer's instructions. The maximum length of the exhaust duct shall be determined by the dryer manufacturer's installation instructions. The code official shall be provided with a copy of the installation instructions for the make and model of the dryer. Where the exhaust duct is to be concealed, the installation instructions shall be provided to the code official prior to the concealment inspection. In the absence of fitting equivalent length calculations from the clothes dryer manufacturer, Table 504.6.4.1 shall be used.

504.6.5 Length identification. When the exhaust duct is concealed within the building construction, the equivalent length of the exhaust duct shall be identified on a permanent label or tag. The label or tag shall be located within 6 feet (1829 mm) of the exhaust duct connection.

IMC § 505 DOMESTIC KITCHEN EXHAUST EQUIPMENT
IMC § 505.1 Domestic systems
Where domestic range hoods and domestic appliances equipped with downdraft exhaust are located within dwelling units, such hoods and appliances shall discharge to the outdoors through sheet metal ducts constructed of galvanized steel, stainless steel, aluminum, or copper. Such ducts shall have smooth inner walls and shall be air tight and equipped with a backdraft damper.

Exceptions:

1. *Where installed in accordance with the manufacturer's installation instructions and where mechanical or natural ventilation is otherwise provided in accordance with Chapter 4, listed and labeled ductless range hoods shall not be required to discharge to the outdoors.*

2. *Ducts for domestic kitchen cooking appliances equipped with downdraft exhaust systems shall be permitted to be constructed of Schedule 40 PVC pipe and fittings provided that the installation complies with all of the following:*
   2.1. *The duct shall be installed under a concrete slab poured on grade.*
   2.2. *The under-floor trench in which the duct is installed shall be completely backfilled with sand or gravel.*

IMC § 505.2 Make-up air required

Exhaust hood systems capable of exhausting in excess of 400 cfm (0.19 m³/s) shall be provided with make-up air at a rate approximately equal to the exhaust air rate. Such make-up air systems shall be equipped with a means of closure and shall be automatically controlled to start and operate simultaneously with the exhaust system.

IMC § 512 SUBSLAB SOIL EXHAUST SYSTEMS
IMC § 512.1 General

When a subslab soil exhaust system is provided, the duct shall conform to the requirements of this section.

IMC § 512.2 Materials

Subslab soil exhaust system duct material shall be air duct material listed and labeled to the requirements of UL 181 for Class 0 air ducts, or any of the following piping materials that comply with the International Plumbing Code as building sanitary drainage and vent pipe: cast iron; galvanized steel; brass or copper pipe; copper tube of a weight not less than that of copper drainage tube, Type DWV; and plastic piping.

IMC § 512.3 Grade

Exhaust system ducts shall not be trapped and shall have a minimum slope of one-eighth unit vertical in 12 units horizontal (1-percent slope).

IMC § 512.4 Termination

Subslab soil exhaust system ducts shall extend through the roof and terminate at least 6 inches (152 mm) above the roof and at least 10 feet (3048 mm) from any operable openings or air intake.

IMC § 512.5 Identification

Subslab soil exhaust ducts shall be permanently identified within each floor level by means of a tag, stencil, or other approved marking.

IMC § 514 ENERGY RECOVERY VENTILATION SYSTEMS
IMC § 514.1 General

Energy recovery ventilation systems shall be installed in accordance with this section. Where required for purposes of energy conservation, energy recovery ventilation systems shall also comply with the International Energy Conservation Code.

IMC § 514.2 Prohibited applications

Energy recovery ventilation systems shall not be used in the following systems:

1. Hazardous exhaust systems covered in Section 510.

2. Dust, stock, and refuse systems that convey explosive or flammable vapors, fumes, or dust.

3. Smoke control systems covered in Section 513.

4. Commercial kitchen exhaust systems serving Type I and Type II hoods.

5. Clothes dryer exhaust systems covered in Section 504.

IMC § 514.3 Access

A means of access shall be provided to the heat exchanger and other components of the system as required for service, maintenance, repair, or replacement.

IMC § 601 DUCT SYSTEMS GENERAL
IMC § 601.1 Scope

Duct systems used for the movement of air in air-conditioning, heating, ventilating, and exhaust systems shall conform to the provisions of this chapter except as otherwise specified in Chapters 5 and 7.

Exception: Ducts discharging combustible material directly into any combustion chamber shall conform to the requirements of NFPA 82.

[B]IMC § 601.2 Air movement in egress elements

Corridors shall not serve as supply, return, exhaust, relief, or ventilation air ducts.
Exceptions:

1. *Use of a corridor as a source of make-up air for exhaust systems in rooms that open directly onto such corridors, including toilet rooms, bathrooms, dressing rooms, smoking lounges, and janitor closets, shall be permitted, provided that each such corridor is directly supplied with outdoor air at a rate greater than the rate of make-up air taken from the corridor.*

2. *Where located within a dwelling unit, the use of corridors for conveying return air shall not be prohibited.*

3. *Where located within tenant spaces of 1000 square feet (93 m^2) or less in area, utilization of corridors for conveying return air is permitted.*

[B]IMC § 601.3 Exits

Equipment and ductwork for exit enclosure ventilation shall comply with one of the following items:

1. Such equipment and ductwork shall be located exterior to the building and shall be directly connected to the exit enclosure by ductwork enclosed in construction as required by the International Building Code for shafts.

2. Where such equipment and ductwork is located within the exit enclosure, the intake air shall be taken directly from the outdoors and the exhaust air shall be discharged directly to the outdoors, or such air shall be conveyed through ducts enclosed in construction as required by the International Building Code for shafts.

3. Where located within the building, such equipment and ductwork shall be separated from the remainder of the building, including other mechanical equipment, with construction as required by the International Building Code for shafts.

In each case, openings into fire-resistance-rated construction shall be limited to those needed for maintenance and operation and shall be protected by self-closing fire-resistance-rated devices in accordance with the International Building Code for enclosure wall opening protectives. Exit enclosure ventilation systems shall be independent of other building ventilation systems.

IMC § 601.4 Contamination prevention

Exhaust ducts under positive pressure, chimneys, and vents shall not extend into or pass through ducts or plenums.

IMC § 602 PLENUMS

IMC § 602.1 General

Supply, return, exhaust, relief, and ventilation air plenums shall be limited to uninhabited crawl spaces, areas above a ceiling or below the floor, attic spaces, and mechanical equipment rooms. Plenums shall be limited to one fire area. Fuel-fired appliances shall not be installed within a plenum.

IMC § 602.2 Construction

Plenum enclosures shall be constructed of materials permitted for the type of construction classification of the building.

The use of gypsum boards to form plenums shall be limited to systems where the air temperatures do not exceed 125°F (52°C) and the building and mechanical system design conditions are such that the gypsum board surface temperature will be maintained above the airstream dew-point temperature. Air plenums formed by gypsum boards shall not be incorporated in air-handling systems utilizing evaporative coolers.

IMC § 602.2.1 Materials within plenums

Except as required by Sections 602.2.1.1 through 602.2.1.5, materials within plenums shall be noncombustible or shall have a flame spread index of not more than 25 and a smoke-developed index of not more than 50 when tested in accordance with ASTM E 84 or UL 723.

Exceptions:

1. *Rigid and flexible ducts and connectors shall conform to Section 603.*

2. *Duct coverings, linings, tape, and connectors shall conform to Sections 603 and 604.*

3. *This section shall not apply to materials exposed within plenums in one- and two-family dwellings.*

4. *This section shall not apply to smoke detectors.*

5. *Combustible materials fully enclosed within noncombustible raceways or enclosures, approved gypsum board assemblies, or enclosed in materials listed and labeled for such application.*

IMC § 603.9 Joints, seams, and connections

All longitudinal and transverse joints, seams, and connections in metallic and nonmetallic ducts shall be constructed as specified in SMACNA HVAC Duct Construction Standards—Metal and Flexible and NAIMA Fibrous Glass Duct Construction Standards. All joints, longitudinal and transverse seams, and connections in ductwork shall be securely fastened and sealed with welds, gaskets, mastics (adhesives), mastic-plus-embedded-fabric systems, or tapes. Tapes and mastics used to seal ductwork listed and labeled in accordance with UL 181A shall be marked "181A-P" for pressure-sensitive tape, "181 A-M" for mastic, or "181 A-H" for heat-sensitive tape. Closure systems used to seal flexible air ducts and flexible air connectors shall comply with UL 181B and shall be marked "181B-FX" for pressure-sensitive tape or "181B-M" for mastic. Duct connections to flanges of air distribution system equipment shall be sealed and mechanically fastened. Mechanical fasteners for use with flexible nonmetallic air ducts shall comply with UL 181B and shall be marked "181B-C." Closure systems used to seal metal ductwork shall be installed in accordance with the manufacturer's instructions. Unlisted duct tape is not permitted as a sealant on any metal ducts.

Exception: Continuously welded and locking-type longitudinal joints and seams in ducts operating at static pressures less than 2 inches of water column (500 Pa) pressure classification shall not require additional closure systems.

IMC § 605 AIR FILTERS
IMC § 605.1 General

Heating and air-conditioning systems of the central type shall be provided with approved air filters. Filters shall be installed in the return air system, upstream from any heat exchanger or coil, in an approved convenient location. Liquid adhesive coatings used on filters shall have a flash point not lower than 325°F (163°C).

IMC § 605.2 Approval

Media-type and electrostatic-type air filters shall be listed and labeled. Media-type air filters shall comply with UL 900. High efficiency particulate air filters shall comply with UL 586. Electrostatic-type air filters shall comply with UL 867. Air filters utilized within dwelling units shall be designed for the intended application and shall not be required to be listed and labeled.

IMC § 605.3 Airflow over the filter

Ducts shall be constructed to allow an even distribution of air over the entire filter.

TABLE 405.5.2(1) Specifications for the standard reference and proposed designs.

Building Component	Standard Reference Design	Proposed Design
Air exhance rate	Specific leakage area (SLA)[d] = 0.00036 assuming no energy recovery	For residences that are not tested, the same as the standard reference design. For residences without mechanical ventilation that are tested in accordance with ASHRAE 119, Section 5.1, the measured air exchange rate[e] but not less than 0.35 ACH For residences with mechanical ventilation that are tested in accordance with ASHRAE 119, Section 5.1, the measured air exchange rate[e] combined with the mechanical ventilation rate[f], which shall not be less than $0.01 \times CFA + 7.5 \times (Nbr+1)$ where: CFA = conditioned floor area Nbr = number of bedrooms
Mechanical Ventilation	None, except where mechanical ventilation is specified by the proposed design, in which case: Annual vent fan energy use: kWh/yr = $0.03942 \times CFA \times 29.565 \times (Nbr+1)$ where: CFA = conditioned floor area Nbr = number of bedrooms	As proposed

d. Where leakage area (L) is defined in accordance with Section 5.1 of ASHRAE 119 and where: SLA = L/CFA where L and CFA are in the same units.

e. Tested envelope leakage shall be determined and documented by an independent party approved by the code official. Hourly calculations as specified in the 2001 ASHRAE Handbook of Fundamentals, Chapter 26, page 26.21, Equation 40 (Sherman-Grimsrud model) or the equivalent shall be used to determine the energy loads resulting from infiltration.

f. The combined air exchange rate for infiltration and mechanical ventilation shall be determined in accordance with Equation 43 of 2001 ASHRAE Handbook of Fundamentals, page 26.24 and the "whole-house ventilation" provisions of 2001 ASHRAE Handbook of Fundamentals, page 26.19 for intermittent mechanical ventilation.

2009 INTERNATIONAL ENERGY CONSERVATION CODE

IECC § 403.5 Mechanical ventilation

Outdoor air intakes and exhausts shall have automatic or gravity dampers that close when the ventilation system is not operating.

2009 INTERNATIONAL BUILDING CODE

IBC § 1203 VENTILATION

1203.1 General

Buildings shall be provided with natural ventilation in accordance with Section 1203.4, or mechanical ventilation in accordance with the International Mechanical Code.

IBC § 1203.2 Attic spaces

Enclosed attics and enclosed rafter spaces formed where ceilings are applied directly to the underside of roof framing members shall have cross ventilation for each separate space by ventilating openings protected against the entrance of rain and snow. Blocking and bridging shall be arranged so as not to inter-fere with the movement of air. A minimum of 1 inch (25 mm) of airspace shall be provided between the insulation and the roof sheathing. The net free ven-tilating area shall not be less than $\frac{1}{300}$ of the area of the space ventilated, with 50 percent of the required ventilating area provided by ventilators located in the upper portion of the space to be ventilated at least 3 feet (914 mm) above eave or cornice vents, with the balance of the required ventilation provided by eave or cornice vents.

IBC § 1203.2.1 Openings into attic.
Exterior openings into the attic space of any building intended for human occupancy shall be protected to prevent the entry of birds, squirrels, rodents, snakes, and other similar creatures. Openings for ventilation having a least a dimension of $\frac{1}{16}$ inch (1.6 mm) min-imum and $\frac{1}{4}$ inch (6.4 mm) maximum shall be permitted. Openings for venti-lation having a least dimension larger than $\frac{1}{4}$ inch (6.4 mm) shall be provided with corrosion-resistant wire cloth screening, hardware cloth, perforated vinyl, or similar material with openings having at least a dimension of $\frac{1}{16}$ inch (1.6 mm) minimum and $\frac{1}{4}$ inch (6.4 mm) maximum. Where combustion air is obtained from an attic area, it shall be in accordance with Chapter 7 of the International Mechanical Code.

IBC § 1203.3 Under-floor ventilation

The space between the bottom of the floor joists and the earth under any building except spaces occupied by a basement or cellar shall be provided with ventilation openings through foundation walls or exterior walls. Such openings shall be placed so as to provide cross ventilation of the under-floor space.

IBC § 1203.4 Natural ventilation

Natural ventilation of an occupied space shall be through windows, doors, louvers, or other openings to the outdoors. The operating mechanism for such openings shall be provided with ready access so that the openings are readily controllable by the building occupants.

IBC § 1203.4.1 Ventilation area required

The minimum openable area to the outdoors shall be 4 percent of the floor area being ventilated.

IBC § 1203.4.1.1 Adjoining spaces

Where rooms and spaces without openings to the outdoors are ventilated through an adjoining room, the opening to the adjoining room shall be unobstructed and shall have an area of not less than 8 percent of the floor area of the interior room or space, but not less than 25 square feet (2.3 m²). The minimum openable area to the outdoors shall be based on the total floor area being ventilated.

Exception: Exterior openings required for ventilation shall be permitted to open into a thermally isolated sunroom addition or patio cover provided that the openable area between the sunroom addition or patio cover and the interior room shall have an area of not less than 8 percent of the floor area of the interior room or space, but not less than 20 square feet (1.86 m²). The minimum openable area to the outdoors shall be based on the total floor area being ventilated.

2009 INTERNATIONAL RESIDENTIAL CODE

IRC § R309 GARAGES AND CARPORTS[5]

IRC § SECTION R408 UNDER-FLOOR SPACE

IRC § R408.1 Ventilation. The under-floor space between the bottom of the floor joists and the earth under any building (except space occupied by a basement) shall have ventilation openings through foundation walls or exterior walls. The minimum net area of ventilation openings shall not be less than

---

5. This section has changed significantly from the 2006 version of the code. Provisions that required sealing the openings from a private garage directly into a room used for sleeping purposes were dropped.

1 square foot (0.0929 m²) for each 150 square feet (14 m²) of under-floor space area, unless the ground surface is covered by a Class 1 vapor retarder material. When a Class 1 vapor retarder material is used, the minimum net area of ventilation openings shall not be less than 1 square foot (0.0929 m²) for each 1500 square feet (140 m²) of under-floor space area. One such ventilating opening shall be within 3 feet (914 mm) of each corner of the building.

IRC § R408.2 Openings for under-floor ventilation. The minimum net area of ventilation openings shall not be less than 1 square foot (0.0929 m²) for each 150 square feet (14 m²) of under-floor area. One ventilating opening shall be within 3 feet (915 mm) of each corner of the building. Ventilation openings shall be covered for their height and width with any of the following materials provided that the least dimension of the covering shall not exceed ¼ inch (6.4 mm):

1. Perforated sheet metal plates not less than 0.070 inch (1.8 mm) thick.

2. Expanded sheet metal plates not less than 0.047 inch (1.2 mm) thick.

3. Cast-iron grill or grating.

4. Extruded load-bearing brick vents.

5. Hardware cloth of 0.035 inch (0.89mm) wire or heavier.

6. Corrosion-resistant wire mesh, with the least dimension being ⅛ inch (3.2 mm).

Exception: The total area of ventilation openings shall be permitted to be reduced to ¹⁄₁₅₀₀ of the under-floor area where the ground surface is covered with an approved Class 1 vapor retarder material and the required openings are placed to provide cross ventilation of the space. The installation of operable louvers shall not be prohibited.

IRC § R408.3 Unvented crawl space. Ventilation openings in under-floor spaces specified in Sections R408. 1 and R408.2 shall not be required where:

1. Exposed earth is covered with a continuous vapor retarder. Joints of the vapor retarder shall overlap by 6 inches (152 mm) and shall be sealed or taped. The edges of the vapor retarder shall extend at least 6 inches (152 mm) up the stem wall and shall be attached and sealed to the stem wall; and

2. One of the following is provided for the under-floor space:

   2.1. Continuously operated mechanical exhaust ventilation at a rate equal to 1 cubic foot per minute (0.47 L/s) for each 50 ft² (4.7 m²) of crawl space floor area, including an air pathway to the common area (such as a duct or transfer grille), and perimeter walls insulated in accordance with Section N1 102.2.9;

2.2. Conditioned air supply sized to deliver at a rate equal to 1 cubic foot per minute (0.47 L/s) for each 50 ft^2 (4.7 m^2) of under-floor area, including a return air pathway to the common area (such as a duct or transfer grille), and perimeter walls insulated in accordance with Section N1 102.2.9;

2.3. Plenum complying with Section M1601.5, if under-floor space is used as a plenum.

IRC § R806 ROOF VENTILATION[6]

IRC § R806.1 Ventilation required. Enclosed attics and enclosed rafter spaces formed where ceilings are applied directly to the underside of roof rafters shall have cross ventilation for each separate space by ventilating openings protected against the entrance of rain or snow. Ventilation openings shall have a least dimension of ¹⁄₁₆ inch (1.6 mm) minimum and ¼ inch (6.4 mm) maximum. Ventilation openings having a least dimension larger than ¼ inch (6.4 mm) shall be provided with corrosion resistant wire cloth screening, hardware cloth, or similar material with openings having a least dimension of ¹⁄₁₆ inch (1.6 mm) minimum and ¼ inch (6.4 mm) maximum. Openings in roof framing members shall conform to the requirements of Section R802.7.

IRC § R806.2 Minimum area. The total net free ventilating area shall not be less than ¹⁄₁₅₀ of the area of the space ventilated except that reduction of the total area to ¹⁄₃₀₀ is permitted provided that at least 50 percent and not more than 80 percent of the required ventilating area is provided by ventilators located in the upper portion of the space to be ventilated at least 3 feet (914 mm) above the eave or cornice vents with the balance of the required ventilation provided by eave or cornice vents. As an alternative, the net free cross-ventilation area may be reduced to ¹⁄₃₀₀ when a Class I or II vapor barrier is installed on the warm-in-winter side of the ceiling.

IRC § R806.3 Vent and insulation clearance. Where eave or cornice vents are installed, insulation shall not block the free flow of air. A minimum of 1-inch (25 mm) space shall be provided between the insulation and the roof sheathing and at the location of the vent.

IRC § M1501 EXHAUST SYSTEMS

IRC § M1501.1 Outdoor discharge. The air removed by every mechanical exhaust system shall be discharged to the outdoors. Air shall not be exhausted into an attic, soffit, ridge vent, or crawl space.

Exception: Whole-house ventilation-type attic fans that discharge into the attic space of dwelling units having private attics shall be permitted.

---

6. See IRC § R806.4 for a description of "Unvented Attic Assembly" requirements.

IRC § M1502 CLOTHES DRYER EXHAUST

IRC § M1502.1 General. Clothes dryers shall be exhausted in accordance with the manufacturer's instructions.

IRC § M1502.2 Independent exhaust systems. Dryer exhaust systems shall be independent of all other systems, and shall convey the moisture to the outdoors.

Exception: This section shall not apply to listed and labeled condensing (ductless) clothes dryers.

IRC § M1502.3 Duct termination

Exhaust ducts shall terminate on the outside of the building. Exhaust duct terminations shall be in accordance with the dryer manufacturer's installation instructions. If the manufacturer's instructions do not specify a termination location, the exhaust duct shall terminate not less than 3 feet (914 mm) in any direction from openings into buildings. Exhaust duct terminations shall be equipped with a backdraft damper. Screens shall not be installed at the duct termination.

IRC § M1502.4 Dryer Exhaust ducts

Dryer exhaust ducts shall conform to the requirements of Sections M1502.4.1 through M1502.4.6.[7]

IRC § M1503 RANGE HOODS
IRC § M1503.1 General

Range hoods shall discharge to the outdoors through a single-wall duct. The duct serving the hood shall have a smooth interior surface, shall be air tight and shall be equipped with a backdraft damper. Ducts serving range hoods shall not terminate in an attic or crawl space or areas inside the building.

Exception: Where installed in accordance with the manufacturer's installation instructions, and where mechanical or natural ventilation is otherwise provided, listed and labeled ductless range hoods shall not be required to discharge to the outdoors.

IRC § M1505 OVERHEAD EXHAUST HOODS
IRC § M1505.1 General

Domestic open-top broiler units shall have a metal exhaust hood, having a minimum thickness of 0.0157-inch (0.3950 mm) (No. 28 gage) with ¼ inch

---

7. The provisions of this section are the same or very similar to the provisions in the International Mechanical Code cited above in section 504. One significant difference is in the length of ducting allowed—25 feet in the residential code and 35 feet in the mechanical code.

(6.4 mm) clearance between the hood and the underside of combustible material or cabinets. A clearance of at least 24 inches (610 mm) shall be maintained between the cooking surface and the combustible material or cabinet. The hood shall be at least as wide as the broiler unit extend over the entire unit, discharge to the outdoors and be equipped with a backdraft damper or other means to control infiltration/exfiltration when not in operation. Broiler units incorporating an integral exhaust system, and listed and labeled for use without an exhaust hood, need not be provided with an exhaust hood.

IRC § M1507 MECHANICAL VENTILATION
IRC § M1507.1 General
Where toilet rooms and bathrooms are mechanically ventilated, the ventilation equipment shall be installed in accordance with this section.
IRC § M1507.2 Recirculation of air
Exhaust air from bathrooms and toilet rooms shall not be recirculated within a residence or to another dwelling unit and shall be exhausted directly to the outdoors. Exhaust air from bathrooms and toilet rooms shall not discharge into an attic, crawl space, or other areas inside the building.
IRC § M1507.3 Ventilation rate
Ventilation systems shall be designed to have the capacity to exhaust the minimum airflow rate determined in accordance with Table M1507.3.

TABLE M1507.3   Minimum required rates for one- and two-family dwellings.

Area To Be Vented	Ventilation Rates
Kitchens	100 cfm intermittent or 25 cfm continuous
Bathrooms—toilet rooms	Mechanical exhaust capacity of 50 cfm intermittent or 20 cfm continuous

For SI: 1 cubic foot per minute = 0.0004719 m^3/s.

2009 INTERNATIONAL PROPERTY MAINTENANCE CODE

IPMC § 303.13.2: Openable windows. Every window, other than a fixed window, shall be easily openable and capable of being held in position by window hardware.

IPMC § 403.1: Ventilation/habitable space. Every habitable space must have at least one openable window. The total openable area of the window in every room shall be equal to at least 45 percent of the minimum glazed area required in Section 402.1.

Exception: Where rooms and spaces without openings to the outdoors are ventilated through an adjoining room, the unobstructed opening to the adjoining room shall be at least 8 percent of the floor area of the interior room or space, but not less than 25 square feet (2.33 m²). The ventilation openings to the outdoors shall be based on a total floor area being ventilated.

IPMC § 403.2 Bathrooms and toilet rooms. Every bathroom and toilet room shall comply with the ventilation requirements for habitable spaces as required by Section 403.1, except that a window shall not be required in such spaces equipped with a mechanical ventilation system. Air exhausted by a mechanical ventilation system from a bathroom or toilet room shall discharge to the outdoors and shall not be recirculated.

STATE VENTILATION CODES

Some states have individual, specific ventilation codes often integrated with their energy efficiency programs. This is by no means a complete list and because these codes are changing regularly, what is listed here may not be the most up-to-date standard. It is important to check with local code officials.

California: California has adopted ASHRAE 62.2-2007 for its ventilation requirements. "If the air tightness of the home is measured and is found to be below 1.5 SLA, mechanical supply ventilation is required." (SLA is the specific leakage area which is the effective leakage area [ELA] divided by the floor area.) "If the builder intends to comply with Title 24 through the performance standards approach, then the total power consumption of any mechanical ventilation must be included as part of the total energy analysis for the home." The primary ventilation system must provide 0.047 cfm/square foot of house conditioned area per the following table:

Square Footage of House	Total Ventilation in CFM	Minimum Duct Diameter
Up to 1300	60	4″
1300 to 1900	90	5″
1900 to 2600	120	6″
2600 to 3200	150	6″
3200 to 3800	180	7″
3800 to 4500	210	7″
4500 to 5100	240	8″

"The preferred method for continuous mechanical ventilation is for the fan to be installed to push air into the house—called the 'supply-fan approach'."

Florida: (The 2007 Florida Building Code is available for free on the ICC site.[8]) The Florida State Building Code Part IV, Chapter 11 Energy Efficiency includes several sections on ventilation including Section N1109.ABC.1 "Buildings operated at positive indoor pressure." It states, "Residential buildings designed to be operated at a positive indoor pressure or for mechanical ventilation shall meet," the design air change per hour minimums in ASHRAE 62, that "no ventilation or air-conditioning system make-up air shall be provided to conditioned spaces from attics, crawl spaces, attached enclosed garages or outdoor spaces adjacent to swimming pools or spas." In a hot, humid climate such as Florida, positive pressure ventilation can prevent that humidity from entering the home through cracks and leaks in the home because the structure will be slightly "inflated."

Minnesota: Minnesota amended the requirements of the International Mechanical Code to comply with the requirements of ASHRAE 62-2001. The ventilation system "shall be designed to supply the required rate of ventilation air continuously during the period the building is occupied, except as otherwise stated in other provisions of the code. 1346.0404 Section 404 Garages amends IMC 404.1 to require, "Mechanical ventilation systems for enclosed parking garages shall provide a minimum exhaust rate of 0.75 cfm per square foot (0.0038 m³/s) of floor area." It does not define whether this is both commercial and residential garages.

There is a lengthy description of make-up air systems and pressure equalization. For example: 501.4.3 paragraph 4 states, "When an exhaust system with a rated capacity greater than 300 cfm (0.144 m³/s) is installed in a dwelling constructed during or after 1994 under the Minnesota Energy Code, Minnesota Rules, chapter 7670, make-up air quantity shall be determined by using IMC Table 501.4.3(1) and shall be supplied according to IMC Section 501.4.2. *Exception: If powered make-up air is electrically interlocked and matched to the airflow of the exhaust system, additional make-up air is not required." The table requires 135 cfm of make-up air for clothes dryers under all conditions.*

Vermont: Revisions to the Vermont Residential Building Energy Code that took effect January 1, 2005, required all newly constructed homes to be

8. ICCsafe Ecodes, Florida.

mechanically ventilated. "Every home must have a system consisting of fans, controls, and ducts that provides the fresh air for the dwelling unit." The whole house or primary ventilation system must be capable of supplying "the specified amount of air during all periods of occupancy automatically without the need for anyone to turn it on or off."[9] The handbook has descriptions of basic ventilation terms and criteria such as sones and building pressures.

State of Washington: Washington has a detailed ventilation code that can be found on the Washington State Building Code Web page under "Ventilation Code."[10] Washington's codes have provisions for installed system testing to verify performance, it limits fan noise for "whole house ventilation systems" to 1.5 sones, and states that the "whole house ventilation fan shall be controlled by a 24-hour clock timer with the capability of continuous operation, manual, and automatic control." The code includes a table for ventilation rates from homes from 500 square feet to more than 9000 square feet and up to eight bedrooms. It also includes a "Prescriptive Exhaust Duct Sizing" table.

SAFETY TESTING AND PERFORMANCE CERTIFICATION

Products are safety tested by a variety of organizations. None of these organizations are government controlled. Most are privately held, not-for-profit underwriting companies. HVI and AMCA are performance certification organizations, verifying to the consumer that similar products will perform in similar, certified ways. Since there are no government agencies that control manufacturers' claims of performance, organizations like HVI and AMCA are vital in assuring the consumer that products with these marks will indeed perform as advertised. Some government programs like Energy Star and California's Title 20 require performance as well as safety testing by a recognized laboratory.

Home Ventilating Institute (HVI)[11]: HVI "represents a wide range of home ventilating products manufactured by companies in the United States, Canada, Asia, and Europe, producing the majority of residential ventilation

9. Vermont Residential Building Energy Code Handbook,
    http://publicservice.vermont.gov/energy-efficiency/ee_resbuildingstandards.html.
10. http://sbcc.wa.gov/Page.aspx?nid=14.
11. http://www.hvi.org.

products sold in North America." HVI's Certified Rating Program was created to provide a credible, third party means for comparing the performance of similar products. "Not only are products certified, but a random verification program ensures that those products still meet their original performance." Performance certification is required for Energy Star recognition.

Air Movement and Control Association International, Inc. (AMCA)[12]: AMCA is a not-for-profit international association of related air system equipment manufacturers, primarily commercial systems. Its Air Movement Division certifies the performance of residential ventilation products.

Underwriter Laboratories (UL) is the best-recognized U.S. product safety underwriter, testing a very wide variety of products. "UL is an independent product safety certification organization that has been testing products and writing standards for safety for over a century. UL evaluates more than 19,000 types of products, components, materials, and systems annually with 21 billion UL marks appearing on 72,000 manufacturers' products each year." UL has 62 laboratories and certification facilities serving customers in 99 countries.

A product that has a UL listing has been tested as a whole. It is a standalone product. A product that has been UL recognized is used as a component of another product. "A product is UL listed if the UL listing mark is on the product, accompanied by the manufacturer's name, trade name, trademark, or other authorized identification."[13] Further, "The UL listing mark on a product is the manufacturer's representation that the complete product has been tested by UL to nationally recognized safety standards and found to be free from reasonably foreseeable risk of fire, electric shock, and related hazards and that the product was manufactured under UL's Follow-Up Services program."

The UL recognized mark means that the component alone meets the requirements for a "limited, specified use."

UL Standard 705 for "power ventilators" has a limited impact on residential "fans intended for heated and conditioned air and for connection to permanently installed wiring systems in accordance with the National Electrical Code, NFPA 70." Alternatively UL Standard 507 covers a high percentage of residential fan products. Products are subjected to a thorough array of safety checks from the flammability of any plastics to electrical safety to the design of the labels.

12. http://www.amca.org.
13. http://www.ul.com/marks_labels/mark/differencev6n1.htm.

Once a manufacturer begins producing the product with the UL listing, UL inspectors regularly visit the manufacturing facility to verify that the product continues to be manufactured as it was initially submitted to UL for testing and that the materials used are the same coming from the same suppliers. If changes are made to materials or suppliers, the manufacturer is required to notify UL of those changes and have the alternatives added to its "Follow-up" procedure.

The addition of a "C" in the UL logo circle indicates that the product has been tested to both U.S. and Canadian safety standards.

Canadian Standards Association International (CSA)[14]: CSA provides a similar and competing testing service. UL and CSA have "harmonized" many of their standards to make them less onerous to manufacturers wishing to sell products on both sides of the border. CSA has a similar follow-up service.

Intertek (ETL)[15]: Intertek (using the ETL mark) "specializes in electrical product safety testing, EMC testing, and benchmark performance testing." It has more than 70 offices and laboratories on six continents. Intertek provides a wide variety of safety testing and other services.

Maryland Electrical Testing (MET) Laboratories[16]: MET Laboratories is a product testing and safety laboratory. They have been certifying, listing, and labeling products for electrical product safety for more than 40 years. They test residential fan products under the UL 507 Standard, and they require a follow-up service. The MET circle mark indicates if the product has been tested to both Canadian and U.S. standards.

14. http://www.csa-international.org.
15. http://www.intertek-etlsemko.com.
16. http://www.metlabs.com.

Chapter 13
PROGRAM REQUIREMENTS AND OPPORTUNITIES

Residential ventilation is a single component of residential building science, and although there are specific skills, knowledge, and expertise that can be applied, the fact is that the "house is a system." There are few educational programs that concentrate solely on residential ventilation. On the one hand that is a bad thing because there is a lot to know. On the other hand it is a good thing because the science of residential ventilation must be a part of the building as a whole.

The strategy should be the same whether you are working on a new or existing building: Build a tight house and ventilate it. The space must be comfortable, safe, and healthy. It must be designed to last a long time, keeping the water and the weather out. It must be serviceable and maintainable or nature will reclaim the space.

There are programs for a doctorate in building sciences that take years to complete. There are conferences in building science that last for four days. There are week-long courses in residential ventilation.

For example, a master of science in building technology at the Massachusetts Institute of Technology (MIT) requires a background in mathematics, physics, and "other technical subjects."[1] Research is done in building control and diagnostics, building energy studies and lighting analysis, building materials and construction, computer graphics and physical performance, indoor air quality, building ventilation, building environment modeling, and sustainable building design.

Many of the degree programs in building science fall into the architecture or civil engineering departments. University of Southern California (USC) describes an applicant for its program as an "individual (who) probably pictures himself as becoming the master builder—the person who builds the structures which are symbols of our civilization."[2]

1. MIT Website, http://architecture.mit.edu/masters-smbt.html.
2. http://www.usc.edu/dept/civil_eng/dept/admission/undergraduate/bldgscience.htm.

The basic science of structures is well known. Craftsmen learned why buildings stand up or why they fall down many years ago. Populations have learned how to create homes that are remarkably well adapted to their environments—from cliff dwellers to adobe structures to igloos to thatched homes to New England saltboxes and high tech modulars. As we have improved the indoor environment by sealing ourselves off from the outdoor environment, the amount of technical knowledge required has increased dramatically. When breezes blew through wall structures, they got wet and then dried continuously. Now they may stay wet. Materials and tools have changed, making it almost impossible to build a house that is loose enough to provide adequate, natural ventilation. Codes have demanded that homes be constructed better. Economics and concerns about the long-term conditions on the earth have demanded that energy loads be reduced.

Many of these conditions bring out the opportunists that will take advantage of popular concerns with "snake-oil" solutions that at best don't really work and at worst do significant harm. Honest knowledge is the best way to avoid these problems.

The knowledge base is changing everyday. Despite the fact that we have been building homes for thousands of years, the new conditions and technologies have given us tools that we can use to measure building performance, "look" into existing walls, "see" building pressure, temperature, and humidity differences, and test houses and ducting for air leaks. These tools and technologies are only a few decades old at most. We are just learning how to use them effectively and what the outcomes mean.

CONTRACTOR TRAINING AND CERTIFICATION PROGRAMS

Because this is a relatively new field, the certification programs are confusing and organizations are still jockeying for position. Contractors have always been locally controlled, and building science professionals continue in that vein. All of the renewed interest in energy saving and the cost of conditioning has explosively affected the industry and the training and knowledge base. Consequently, the information included here will be changing, but the resources should serve as a place to start.

Two primary building science organizations in the U.S. are RESNET[3] (Residential Energy Services Network), which operates the HERS program

---

3. http://www.natresnet.org.

(Home Energy Rating System) and BPI[4] (Building Performance Institute). Residential ventilation is a component but not the focus of their programs. Many other organizations use HERS or BPI ratings as templates or integral parts of their own programs.

There is one primary resource for learning about residential ventilation in the U.S. or Canada. The Heating, Refrigeration, and Air Conditioning Institute's (HRAI) (the Canadian national HVACR industry association) SkillTech Academy[5] is an intensive course on residential ventilation skills. This is an in-depth program that covers the fundamentals of building science that particularly relate to residential ventilation and the details of ducting design, installation, and testing. It provides checklists, charts, sizing, commissioning, and maintenance lists. The course provides an in-depth component for duct design for HRV/ERV systems. These systems are considerably more complex than simply throwing a bathroom fan in the ceiling and venting it to the outside. The duct design for an independent HRV/ERV installation is equivalent to the detailed duct design work that is required for a ducted HVAC system. Once it has been installed, the flows should be reasonably balanced between supply and exhaust and require only minor adjustment. Design criteria include the location of the grilles and the connections to the air handler if the two systems are combined. The course also details combustion safety issues concerning depressurization of the building as well as other building pressure issues.[6] HRAI describes their two-day certification program in "residential mechanical ventilation installation" as:

> *The Residential Mechanical Ventilation Installation Course is designed for contractors and building professionals who may be interested in developing opportunities in the growing residential ventilation market, in addition to those currently involved with designing and installing residential mechanical ventilation systems including energy recovery ventilators. The course covers the fundamentals of air quality assessment, system requirements, and focuses specifically on system design and installation. Course participants receive the complete HRAI Residential Mechanical Ventilation Manual, as endorsed by the Home Ventilating Institute (HVI). Those who achieve 75 percent or higher on the exam will obtain HRAI Certification and will be listed on the U.S. Certification List on HRAI's Website.*

4. http://www.bpi.org/content/home/index.php.
5. www.hrai.ca.
6. HRAI also offers an in-depth course on indoor air quality.

The Home Ventilating Institute (HVI)[7] is the U.S. counterpart to HRAI, but (as of this writing), except for sessions at various conferences, they do not offer training programs. HVI's Certified Products Directory is particularly useful in providing third party performance testing of ventilation products. HVI's Website has connections to resources for sizing and specifying systems and provides downloads of the "Fresh Ideas" ventilation guide.

Some product manufacturers including Panasonic offer extensive information on their Websites including Panasonic's Ventilation University.[8] The information is very detailed for design, sales, and installation professionals. Panasonic also offers a useful sizing program[9] to select the right size product to meet the ASHRAE 62.2 Standard for required airflow.

Vent Axia, a British fan company, has one of the most extensive resources for residential ventilation online. Their products and applications apply to the European market, but the information on air and air motion is extremely informative.[10]

North American Technician Excellence (NATE) is an organization that certifies heating, ventilation, air conditioning, and refrigeration technicians. It is an independent, third party, nonprofit certification organization. They say, "NATE tests technicians; others train."[11] Although NATE does not actually supply training, other organizations offer courses that are NATE recognized. The HRAI Residential Mechanical Ventilation Installation course, for example, qualifies for 16 hours of continuing education credits toward NATE recertification.

The Building Performance Institute (BPI)[12] also provides certification of building science knowledge, recognizing other training programs. BPI accredited professionals have had to pass a BPI examination including both classroom and field testing, and they have to maintain that certification through on-going education and training. (The HRAI ventilation course is BPI-recognized and qualifies for 16 hours of continuing education for individuals certified as a BPI heating specialist.)

As of this writing, BPI has more than 50 organizations that offer training on building performance issues that follow BPI education templates. Many of these organizations list and certify professionals under their own banners.

7. www.hvi.org.
8. http://panasonicvu.buildingmedia.com/design.php.
9. http://www2.panasonic.com/webapp/wcs/stores/servlet/vVentFanSelector?storeId= 15001.
10. http://www.vent-axia.com/knowledge/handbook/section1.asp.
11. http://www.natex.org/HVAC_HVACR/about_home.html.
12. http://www.bpi.org/content/home/index.php.

You'll find a listing of many of these in Appendix E. Several of them are described here.

Alaska Housing and Finance Corporation: "Alaska Housing Finance Corporation (AHFC) is a self-supporting public corporation with offices in 16 communities throughout Alaska." Through a wide variety of workshops, AFHC offers training for professionals on building science, the use of blower doors, cold climate retrofits, and air tightness. They also provide workshops for the general public on energy and indoor air quality and to builders, inspectors, lenders, and real estate professionals on home inspections and financing.

Association of Energy Affordability, Inc.: AEA is "committed to using energy efficiency to maintain affordable and healthy housing for low- and moderate-income families and communities." AEA offers BPI template training and certification in weatherization, energy efficient building operations, multi-family building analysis, hydronic heating system design, building analyst, and envelope specialist.

California Building Performance Contractors Association: CBPCA "helps green contractors identify and perform quality green home energy upgrades—our name for comprehensive (i.e., whole house) energy efficiency improvements that improve homeowner comfort and indoor air quality while lowering their energy bills." CBPCA offers HERS training as well as specialized seminars in green home energy upgrades, insulation installation, air sealing, and air conditioning.

The Comfort Institute: "The Comfort Institute is an international indoor comfort research, training, and consumer protection organization based in Bellingham, Washington with trainers and offices located throughout the country." The Comfort Institute provides HVAC training programs.

Conservation Services Group: CSG serves as the training organization for programs in many states provided by many local utilities. They also offer valuable resources by state on their Website.

HVACReducation: This is an online training program. They offer a "GREEN Principles of Building Science" program that was developed and written in partnership with Advanced Energy of North Carolina. The course provides a ventilation component along with the many fundamentals of building science that will help prepare students for NATE, NARI, BPI, and RESNET credentials.

Kansas Building Science Institute: KBSI "was founded in 1996 to provide building science and energy performance training for the weatherization, home energy rating, utility, and building trades." KBSI provides home energy rater certification training accredited by RESNET, as well as BPI analyst training and testing, and special programs in inspecting residential furnaces and mold and moisture.

Saturn Resource Management, Inc.: SRMI delivers online courses for energy professionals. They offer overview courses to "provide training for those just starting in the field of energy auditing" as well as certification and proficiency courses to provide training for career-oriented professionals. They also offer courses for BPI and HERS raters as preparation for taking the certification exams.

Southface Energy Institute: Southface (in Atlanta, GA) offers HERS certification preparation and LEED training events as well as offering its EarthCraft House training and its Homebuilding School and training for contractors to meet the Home Performance with Energy Star guidelines.

Vermont Energy Investment Corporation: "VEIC's mission is to reduce the economic, social, and environmental costs of energy consumption through the promotion of cost-effective energy efficiency and renewable energy technologies." VEIC provides training toward BPI certification for building analysts, heating specialists, and shell specialists.

Wisconsin Energy Conservation Corporation: "WECC works with utilities, municipalities, regulators, government agencies, and consumer groups to meet their energy goals in a verifiable, quantifiable way." WECC is a BPI training affiliate and offers LEED for Homes certifications.

RESNET (Residential Energy Services Network) operates the HERS (Home Energy Rating System) program. This has been mainly for new construction, moving toward a zero energy home, but the auditing technique applies as well to existing buildings. The HERS rater certification program offers building science training that has a ventilation component to it. The program has a thorough quality control element, and detailed records are maintained. Several states are considering a broad application of the HERS rating system.

Community college programs offer incredible opportunities to learn more about more topics than almost any other resource. The programs are too varied and too fluid to describe here, but they are certainly worth looking into.

There are a growing number of online training courses that provide both building science and residential ventilation training. Many equipment manufacturers have expanded their Websites to provide in-depth product selection information (these naturally direct potential customers to their own equipment, but they still provide a substantial amount of useful information). Test equipment manufacturers also provide training in the use of their test equipment. This often occurs at conferences and trade shows.

Some conferences focus on the trade show component and some on the educational aspect of the program. ASHRAE incorporates the largest U.S. trade show for HVAC equipment—AHR Expo (Air-Conditioning, Heating, and Refrigerating). Although it focuses on commercial equipment, there are interesting exhibits that relate to residential applications. The associated conference also has a primarily commercial focus.

More residential building science training can be found at such conferences as the Affordable Comfort Conference, the Energy and Environmental Building Association (EEBA) conference, NESEA (Northeast Sustainable Energy Association) Building Energy Conference, and Southface's Greenprints conference, among others. These conferences feature a body of dedicated, volunteer teachers and building scientists. The conferences often include HERS or BPI training as well as providing hands on experience with equipment, and they are available over a wide geographic area.

GREEN BUILDING PROGRAMS

Green building programs are developing and changing at an astounding rate. What's listed here are just a few of the national and state programs. Virtually all of these programs recognize that when the house is tightened up for energy efficiency that mechanical ventilation is required, and most of them connected their ventilation requirements to ASHRAE 62.2. This listing is provided for a "taste" of what is going on and it is important to do some local research. Building an energy efficient structure makes consummate sense and creates a structure that will be more comfortable to the occupants both physically and economically, and make it more saleable. But it is important to carefully follow the guideline, particularly in regard to mechanical ventilation because as the building gets tighter the margin for error decreases, and there is nothing worse than callbacks and unhappy building owners!

The best-known, national, green building program is administered by the United States Green Building Council (USGBC), LEED for Homes. LEED is a comprehensive program that includes much more than just ventilation. There are eight different areas of environmental quality (EQ) of which ventilation is just one.

The EQ requirements can be met by following the steps for Energy Star with Indoor Air Plus whose ventilation requirements follow ASHRAE 62.2-2007. Alternatively EQ4 prescribes the requirements for outdoor air ventilation, which also requires ASHRAE 62.2-2007 be followed as a prerequisite or a passive ventilation system that is verified by a licensed HVAC engineer. These systems can be enhanced through the use of a variety of controls or third party testing.

The NAHB has developed a "Green Building Standard" that has been ANSI recognized and can be purchased from ICC as Standard 700-2008. The standard provides points for various ventilation system configurations, although no ventilation system is mandatory.

The Green Communities[13] building criteria for green building serves as an excellent resource for all aspects of green building above and beyond ventilation. They recommend the use of ASHRAE 62.2 and 62.1. They have references to the Energy Star Websites and provide "things to consider" referring to LEED for Homes requirements, HVI references, and the Building Science company's "Residential Ventilation Technologies"[14] which links to a 2005 paper on residential ventilation systems.

One local program, Wisconsin's Green Built Home,[15] includes points for two properly supported ceiling fans, installing heat or energy recovery ventilators, and a high efficiency whole house fan installed with an R-38 insulated cover. The program also awards points for attached garages ventilated to "neutral pressure" and Energy Star rated bath fans.

ASHRAE STANDARDS

ASHRAE (The American Society of Heating, Refrigeration, and Air Conditioning Energy) is the U.S. national professional HVAC association. ASHRAE has developed Standard 62.2 for low-rise residential buildings of three stories or fewer above grade and Standard 62.1 for all other residential structures. ASHRAE does offer online training courses on ventilation and HVAC systems, although its emphasis is on commercial buildings.

ASHRAE offers Guideline 24-2008 to assist in the implementation of the ASHRAE 62.2-2007 Standard. The Guideline also offers extensive amplification of IAQ issues, international thresholds, system design and commissioning. Some of this information is included in the informative appendices at the end of the 62.2-2007 Standard. The standard is in continuous maintenance and ongoing changes and amplifications are made and reflected in addenda.

13. http://www.greencommunitiesonline.org.
14. http://www.buildingscience.com/documents/reports/rr-0502-review-of-residential-ventilation-technologies/view?topic=/doctypes/researchreport
15. http://wi-ei.org/greenbuilt/index.php?category_id=3981.

Standard 62.2-2007 consists of a number of parts including purpose, scope and definitions, but the heart of the standard is the definition of the primary ventilation system or "whole building ventilation" system sizing and design, a description of "local exhaust" systems, and "other requirements" such as "transfer air," system instructions, clothes dryer installation, and garages (among others).

Using the standard is straightforward:

- Select a primary or whole building ventilation system that matches the size of the house and the number of bedrooms;

- Run this system continuously (if intermittent operation is preferable, the system has to be more powerful);

- Add local exhaust to the bathrooms and kitchen;

- Make sure that make-up or transfer air is coming directly from the outside and not from polluted sources like garages and crawl spaces;

- Make sure clothes dryers are vented to the outside and that combustion and fuel burning appliances have adequate combustion and ventilation air;

- Make sure the house/garage interface is carefully sealed and that HVAC systems in the garage have little to no leakage;

- Provide a filter for supply ventilation systems with a minimum efficiency of MERV 6 and ensure that inlets are located at least 10 feet (3 meters) from known contaminant sources and won't be obstructed by snow or plantings;

- Use equipment that has been tested and certified by HVI to HVI 915 for sound (1.0 sone or less for continuously operated fans and 3.0 sones or less for intermittently operating fans) and HVI 916 for airflow;

- Install the equipment so that it performs correctly.

Ideally, according to 62.2-2007, the primary or whole building ventilation system should run continuously at a relatively low rate, changing all the air in the building. The ventilation rate for this system is defined by the table (Table 3.1a and Table 3.1b) or by using the formula described in Chapter 3 of this book. If the system cannot be run continuously, its effectiveness, ϵ, will be impacted.[16] More air will need to be moved over a shorter period of time to accomplish the same effective removal of pollutants. The standard defines the required flow rate with the formula:

$$Q_f = Q_r \div (\epsilon \times f)$$

16. Note that this process is in flux as this is written. Check the ASHRAE site for the latest details.

where

Q_f = fan flow rate

Q_r = ventilation air requirement (from Table 3.1a or 3.1b)

ϵ = ventilation effectiveness (from Table 13.1)

f = fractional on-time (defined as the on-time for one cycle divided by the cycle time)

Fractional run time can prove to be a significant increase in flow rate. For example, a fan that runs 40 percent of the time or 9.6 hours per day (on for 9.6 hours and off for 14.4 hours) would have a ventilation effectiveness of 0.46. In a 1200 square foot house with two bedrooms and a 45 cfm ventilation requirement, the fan would have to be:

$$45/(0.40 \times 0.46) = 245 \text{ cfm}$$

If the control was set up to run the fan 20 minutes each hour, the effectiveness would equal 1 and the required flow would be:

TABLE 13.1[17]   Ventilation effectiveness for intermittent fans. (From ASHRAE 62.2-2007 Table 4.2)

Fractional On-Time	Cycle Time (hours)*			
f	0–6	8	12	24
0.1	1.00	0.87	0.65	**
0.2	1.00	0.90	0.76	**
0.3	1.00	0.93	0.83	**
0.4	1.00	0.95	0.88	0.46
0.5	1.00	0.96	0.92	0.68
0.6	1.00	0.98	0.95	0.81
0.7	1.00	0.99	0.97	0.90
0.8	1.00	0.99	0.99	0.96
0.9	1.00	1.00	1.00	0.99
1.0	1.00	1.00	1.00	1.00

*Fan cycle time, defined as the total time for one on-cycle and one off-cycle;
**Condition not allowed since no amount of intermittent ventilation will provide equivalent indoor air quality.

17. See the latest addenda for clarifications of this process.

$$45/(0.33 \times 1) = 136 \text{ cfm}$$

Standard 62.1-2007 "applies to all spaces intended for human occupancy except those within single-family houses, multi-family structures of three stories or fewer above grade . . ."[18] 62.1-2007 applies to a wide variety of spaces from jails to conference rooms. It does include residential applications, which need to be extracted from the wide variety of pollutant issues that the standard addresses.

This standard allows the use of one of two procedures to determine the airflow rate: the ventilation rate procedure or the IAQ procedure. The ventilation rate procedure utilizes space type/application, occupancy level, and floor area to determine air intake rates. The IAQ Procedure is a design procedure "in which outdoor air intake rates and other system design parameters are based on an analysis of contaminant sources, contaminant concentration targets, and perceived acceptability targets." The majority of residential applications would use the ventilation rate procedure, which determines the required airflow by adding the flow rate per person times the number of people and the flow rate per unit area times the area.

People outdoor air rate—5 cfm per person (2.5 L/s per person)

Area outdoor air rate—0.06 cfm/ft2 (0.3 L/s-m²)

"Default occupancy for dwelling units shall be two persons for studio and one-bedroom units, with one additional person for each additional bedroom.

"Air from one residential dwelling shall not be recirculated or transferred to any other space outside of that dwelling."

The standard also requires spot ventilation from the kitchen (50 cfm [25 L/s] continuous, 100 cfm [50 L/s] intermittent) and bathrooms (25 cfm [12.5 L/s] continuous, 50 cfm [25 L/s] intermittent).

Standard 62.1 includes a great deal of information on indoor air issues and should be directly referred to for application details.

Standard 119-1988—Air Leakage Performance for Detached Single-Family Residential Buildings is to establish the performance requirements designed to reduce the air infiltration load and classify the air tightness of residential buildings.

Standard 136-1993—A Method of Determining Air Change Rates in Detached Dwellings is for use in evaluating the impact of air change range in detached dwellings on indoor air quality.

---

18. ASHRAE Standard 62.2-2007, page 3.

HEALTHY HOUSING PROGRAMS

Health House® is a program developed by the American Lung Association of the Upper Midwest.[19] Health House requirements include the "most stringent building standards in the U.S., which include site inspections during construction and performance testing upon completion." Health House components include advanced framing techniques, foundation waterproofing and moisture control, air sealing and advanced insulation techniques, whole house ventilation, and humidity control, among other issues.

Housing and Urban Development's Healthy Homes Program[20]: This program was launched in 1999 to protect children and their families from housing-related health and safety hazards. "The Healthy Homes Program addresses multiple childhood diseases and injuries in the home. The initiative takes a comprehensive approach to these activities by focusing on housing-related hazards in a coordinated fashion, rather than addressing a single hazard at a time." Ventilation is a component to the program for improvement of indoor air quality, particularly addressing allergens, asthma, carbon monoxide, pesticides, and radon.

Healthy Homes Initiative[21]: The Center for Disease Control (CDC) has developed this initiative to provide a "coordinated, comprehensive, and holistic approach to preventing diseases and injuries that result from housing-related hazards and deficiencies. The focus of the initiative is to identify health, safety, and quality-of-life issues in the home environment and to act systematically to eliminate or mitigate problems."

Healthy Housing Solutions was established by the National Center for Healthy Housing in November 2003 to provide federal, state, and local agencies as well as the private sector organizations with professional services related to residential environmental and safety issues. Healthy Housing Solutions teams up with other organizations such as Affordable Comfort, Inc. to provide training in such areas as lead abatement, ventilation, and mold mitigation.

Healthy House Institute "provides consumers information to make their homes healthier. HHI strives to be the most comprehensive educational resource available for creating healthier homes."[22] HHI provides extensive information on many aspects of healthy homes (including ventilation) on its site.

19. http://www.healthhouse.org.
20. http://www.hud.gov/offices/lead/hhi/index.cfm.
21. http://www.cdc.gov/healthyplaces/healthyhomes.htm.
22. http://www.healthyhouseinstitute.com.

Chapter 14
FAN TYPES AND APPLICATIONS

There are a lot of different fan designs, each suited to particular applications. Most of them are hidden inside their systems like the blowers in the bath fan in the ceiling or the motorized impeller in the in-line fan. Some of them are pretty obvious like the big paddle fan in the ceiling or the oscillating propeller fan, twisting back and forth in the hot summer afternoon air. Fan product designers don't just randomly select a fan design because they happen to like the way it looks. A great deal of engineering thought goes into even the least expensive ventilation product. For a while, residential ventilation products didn't change much. But as homes become tighter and mechanical ventilation becomes a common thread by design and by code, residential ventilation system manufacturers are aggressively competing to produce the quietest and most energy efficient products with the greatest economic value possible. All ventilation fans are designed to move air, but some are better at pushing or pulling air through ducts and some are better at gently stirring up the air in a room. Neither would do the other's job well. Selecting the wrong product for the application means that the job won't get done efficiently and may not work at all.

FAN LAWS

It is useful to have some familiarity with fan laws in designing a ventilation system as they provide a means for predicting the effects of altered operating conditions. The mechanical efficiency of a fan remains constant throughout its range of operating speeds (revolutions per minute or rpm). The performance of fans and their relationship to the system in which they are installed is governed by the principles of fluid dynamics.

Fan law variables are the fan size, the rotational speed, the density of the gas being moved, the volume of air being moved, pressure, power (in watts), and the mechanical efficiency. Fan law one describes the effect of changing the fan size, the rotational speed, or the density of the gas on the cubic feet of air moved, the pressure, and the power. Fan law two describes the effect of changing the fan size, pressure, or density on cubic feet of air moved, the rotational speed, and the power. Fan law three describes the effect of changing the fan size, airflow volume, and gas density on rotational speed, pressure, and power. For all the fan laws, the fans being considered must be aerodynamically similar. The fan laws may also be applied to the same fan at different speeds when all the other flow conditions are the same.[1]

A variation of fan law one says that the system pressure (SP) (or total pressure (TP) or velocity pressure (VP)) will vary directly as the square of the change in rotational speed. In other words:

New SP (or TP or VP) = (New rpm/initial rpm)2 × Initial SP (or TP or VP)

So if the static pressure of the system was 0.15 inch wg and the fan was spinning at 1500 rpm and it dropped to 1200 rpm, the new static pressure would be 0.096 inch wg. An example of fan law two says that the power used will vary directly as the square of the change in fan size. In other words:

$$\text{New power (watts or BHP or AHP)} = \text{Initial power} \times \text{(New fan size/Old fan size)}$$

So if the system was using an axial fan with a 10 inch diameter blade using 50 watts and the fan was replaced with a 12 inch diameter bladed fan of the same type, the new power used would be 72 watts.

In one variation, fan law three says that rotational fan speed will vary directly as the cfm ratio. In other words:

New rpm = Initial rpm × (New cfm/Initial cfm)

So if a fan that was spinning at 1500 rpm moving 500 cfm had its speed decreased and had a new measured airflow of 400 cfm, its new rpm would be 1200. The air volume (cfm) varies directly with the fan speed. The pressures vary as the square of the fan speed. The brake horsepower (BHP) varies as the cube of the fan speed.

For most residential ventilation applications the density of the gas and the mechanical efficiency of the system are not variables of great concern. Most installers will rely on the product measurements made by the manufacturer, but it is interesting to understand why manufacturers chooses the products they choose for various applications. Understanding what happens to the power consumption when the size of the fan increases or how the rpm can be

---

1. See ASHRAE Handbook: HVAC Systems and Equipment for the complete fan law listing.

reduced to quiet the system down helps to find ways to analyze and resolve problems.

AXIAL FANS

The most familiar type of fan is the axial (see Figure 14.1). They are fans that we see all the time in windows, on pedestals, oscillating back and forth in hot summer weather. They are used in cooling computers and other electronic products, any place where they don't have to work too hard to move the air in, under, around, or through some resistance. With an axial flow fan, the air flows through in an axial direction, from one side directly or straight through to the other side, parallel to the shaft of the fan blade. They generally consist of a direct drive or belt drive motor, a hub or central "spider," a series of blades, a venturi housing, and a mounting platform for the motor. In residential applications, the belt driven fans are most commonly found in whole house comfort ventilators mounted in the attic floor or as attic exhaust fans.

The belt drive adds some flexibility to the fan design as the speed of rotation of the fan blade can be varied and high-speed motors are generally less expensive than slower ones of the same horsepower. They need to have their belts checked as a part of regular maintenance and they are less efficient than direct drive systems since the belt drive consumes 10 to 15 percent of the brake horsepower of the system.

FIGURE 14.1   An axial ceiling fan. (Broan)

The direct drive approach puts the fan motor directly in the middle of the air stream behind the hub of the blade where it doesn't restrict the airflow (see Figure 14.2). This makes it a more compact design that is more suitable for restricted space. It also makes it less expensive because there are fewer parts, which also reduces maintenance issues.

The efficiency of airflow delivery depends on optimizing all the elements of the fan design. Sound level, for example, can be reduced by using a bigger blade but rotating it more slowly. The pitch and curvature of the blade, the number of blades relative to the number of motor mounting struts, the prox-

FIGURE 14.2   A small axial fan. (EBM-papst)

imity of the end of the blade to the inside of the venturi all affect efficiency, sound level, and delivered air. The optimum design may not be robust enough for a residential application if all the elements of the delivery of the product from the manufacturer to the retailer to the installer are taken into account. A blade that is perfectly located in the venturi housing for optimum performance may shift just enough to rub the housing after installation, causing the fan to be discarded.

A *propeller* fan is the simplest of axial fans commonly used as a panel fan in a window or a "man cooler" fan, oscillating back and forth. It is generally a direct drive arrangement with the motor in the air stream, a number of fairly large blades, and limited components for smoothing or directing the airflow.[2]

A *tubeaxial* fan is an axial fan in a tube or cylinder. They are known for high airflows at relatively low operating pressures. The efficiency of a tubeaxial fan can be as high as 65 percent. The outside of their housings does not have to be cylindrical. Square tubeaxial fans are the fans found in electronics. Larger tubeaxial fans are used in industrial and commercial applications like paint booths where they may be subjected to contaminated or corrosive materials.

A *vaneaxial* fan is a tubeaxial fan with guide vanes to improve the efficiency by straightening and directing the airflow. The vaneaxial design allows for high performance in a minimal space. The efficiency of vaneaxial fans can be as high as 75 percent. They are used in applications requiring high volume airflows at moderate to high pressures.

Axial fans can be installed in a side-by-side configuration to increase the volume of the air that they move or they can be installed in series, one behind the other to increase their strength, their ability to push the air through a duct. In some configurations, in a series arrangement the blades can be spinning in the same rotational direction and sometimes in opposite directions. When they spin in the same direction, they increase the rotational "twist" of the air through the duct. When they spin in opposing directions, the air stream spin is reversed by the second fan or fan blade and straightened. Stacking the fan blades or mounting them in series on the motor shaft can have somewhat the same strengthening effect, although adjustments have to be made to the fan motor selection to allow the blades to spin at an effective rate.

2. Fan Handbook, McGraw-Hill, 1998 by Frank P Bleier provides extensive information on fan design.

CENTRIFUGAL FANS

Centrifugal fans suck the air into the middle, spin it around, and fling it out centrifugally into the ducting, essentially making the air execute a 90-degree turn. Small centrifugal fans are the most common design for ceiling mounted bath fans. A large percentage of centrifugal fans allow air to enter on only one side (a single inlet), which is a typical arrangement in a ceiling mounted bath fan. Some allow air to enter on both sides (double inlet). The motor spins the wheel in a housing that is shaped to collect and guide the air around the perimeter of the wheel and force it out through a duct opening. In most residential ventilation system designs, this housing is called a "scroll." The scroll is carefully shaped to optimize the air pressure of the air leaving the fan housing and entering the ducting. The velocity of the air at the blade tips is considerably higher than the velocity at the housing outlet. The scroll is a spiral shape governed by a series of proportions starting with a relatively narrow passage to the wide open space at the housing outlet. (See the scroll housing description in "Fan Handbook.")[3]

Centrifugal fans come in six basic versions:

1. Centrifugal fans with airfoil (AF) blades;
2. Centrifugal fans with backward curved (BC) blades;
3. Centrifugal fans with backward inclined (BI) blades;
4. Centrifugal fans with radial-tip (RT) blades;
5. Centrifugal fans with forward-curved (FC) blades;
6. Centrifugal fans with radial blades (RB).

The first of these, the centrifugal fan with an airfoil blade, has the best mechanical efficiency. Each stubby blade shaped like an airplanes wing is located around an open circle, standing between a flat plate and a curved, inlet plate.

For all the centrifugal fans, the air located between the blades and rotating with them is subjected to the centrifugal force. That force has a greater effect on the air movement than the shape of the blades with the exception of the forward curved blades. Because of the centrifugal effect, centrifugal fans produce more static pressure than axial fans of the same wheel diameter and running speed because the centrifugal effect adds that additional force to the air movement. In the airfoil bladed version, the shape of the blades does not real-

3. Fan Handbook, Blier, p. 7.11.

ly serve as an airfoil since there is no "lift." It is simply a backward curved blade with a blunt leading edge that improves the structural strength of the blade and allows it to move smoothly and quietly through the air (see Figure 14.3). Airfoil blades can be 92 percent efficient whereas the typical straight, radial bladed design has an approximate efficiency of 60 percent.[4] The other blade designs fall between these two. The fan system designer has to select the fan blade and housing design that best suits the application for efficiency, sound level, cost of manufacture, and ease of assembly and service.

FIGURE 14.3 A backward curved (BC) motorized impeller. (EBM-papst)

4. Fan Handbook, Blier, p. 7.2.

Forward curved blades deliver considerably more air volume and a higher static pressure than airfoil, backward curved, or backward inclined blades, but at lower efficiencies. Forward curved blades can run at about half the speed needed for a comparably sized blade moving the same volume and static pressure. This makes them particularly suitable to the smaller fans in a ventilation system and for cooling electronic equipment where the slower rotational speed will minimize noise and vibration.

The number of blades is governed by compromising between providing a large enough channel to minimize the resistance to the airflow, but tight enough to effectively guide the air through the blade and out through the outlet. The passages between the multitude of blades in the forward curved centrifugal fans is relatively narrow. The edges of the blades are the principal cause of turbulence in the fan, and turbulence means noise and inefficient air movement.

The height of the blades is governed by efficiency and available space, and the length of the blade may slightly overlap the leading edge of the following blade. The space between the inlet and rotating blade should be as small as possible for best performance. (This is another area where there must be a compromise between fan performance, structural integrity, and product delivery to the installed site. Any shift in a tight clearance will cause the blade to rub on the inlet, which in turn will product unacceptable noise and product rejection.)

The highest relative air velocity occurs at the leading edge of the fan blade. That is the point of most concern for noise issues. Noise is less of a concern for centrifugal fans than for axial fans because the air entering the centrifugal fan has experienced turbulence making the 90-degree turn entering the fan, whereas the axial fan's straight-through configuration introduces the turbulence inside the fan. Some additional turbulence created by the fan blade tips is therefore less noticeable in a centrifugal fan. These are relative observations, of course. As the fan "noise" in room mounted bath fans drops below 1 or even ½ sone, every sound-generating element becomes critical to the design. The beauty of minimizing the sound is that it generally includes optimizing the efficiency.

MOTORIZED IMPELLERS

The difference between a standard fan motor and a motorized impeller is that in a standard fan motor the rotor is internal. In a motorized impeller, the rotor is external, spinning with the fan blade. This means that all the heat from the rotor is dissipated into the air stream, keeping it cooler. Simpler adjustment of the speed of a motorized impeller is possible because of the effective heat dissipation.

Motorized impellers are commonly used in residential ventilation as remote mounted or in-line duct fans (see Figure 14.4). They combine the qualities of tubeaxial fans for pulling air through ducting, the compact efficiency and power of centrifugal blowers, and the unique speed adjusting option of the design. They generally use backward curved blades to optimize their static pressure capabilities, working against the resistance of ducting and fittings like hoods and grilles. They are the workhorse fans that can be applied dragging radon-polluted air out from under slabs or operating quietly enough to be used with kitchen range hoods (see Figure 14.5).

Motorized impellers are available with permanent split capacitor (PSC) motors, electronically commutated motors (ECM), or direct current (DC) motors. Both ECM and DC motors are particularly well suited to speed adjustment.

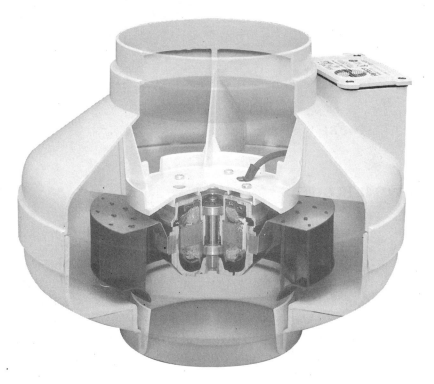

FIGURE 14.4   A motorized impeller in an in-line fan. (Fantech)

FIGURE 14.5 A motorized impeller. (EBM-papst)

TANGENTIAL BLOWERS

In tangential or "cross-flow" blowers the air passes through the forward curved blades of the rotating wheel twice, both on the intake and on the exhaust. Tangential blowers are the long, narrow blowers that are commonly used in devices like copiers or fireplace inserts where the air enters and exits from the blower through long, narrow slots (see Figure 14.6). They can be made to virtually any width desired. The air is induced to make a 90- to almost a 180-degree change in direction. The maximum efficiency of a cross-flow blower is much lower than for a similarly sized, forward curved centrifugal blower because of the air moving twice through the fan wheel and the ensuing rapid changes of direction.

They do produce relatively large volumes of air, but their structural strength is relatively limited, restricting rotational and blade tip speed. And they can be relatively expensive due to manufacturing costs of assembly and balancing.

FIGURE 14.6   A tangential blower. (EBM-papst)

Chapter 15
SPECIAL APPLICATION VENTILATION

The controlled movement of air in a house is not all just for air for the occupants to breathe, and not all air movement in homes happens mechanically. There are areas of particular pollutant concern (if this were a government book that would be PPC) like garages, the radon gas under the slab, crawl spaces, or attics. Clothes dryers and central vacuums are also ventilation systems, but they have their own built-in exhaust fans. It's balancing the make-up air for these devices that you have to be concerned about.

FIREPLACES AND WOODSTOVES

Fireplaces also serve as exhaust systems. They are particularly difficult to deal with because the flow rate up the chimney changes as the heat of the fire changes. The airflow through the chimney is of particular concern at the beginning and the end of the burning cycle. When the fire first starts, the chimney is cold, leading to a poor draft and sometimes causing the fire to smoke or backdraft. At the end of the fire, the chimney cools and the smoldering embers may not generate enough heat to encourage the combustion gases to flow up the chimney. Those incomplete combustion gases include carbon monoxide (CO), which can be deadly. If there are backdrafts at this point in the combustion cycle, the effects can be extremely serious for the health of the occupants.

Fireplace insert manufacturers have come a long way in improving the doors and the combustion chambers, but none of these systems are completely airtight. (Note that many of them will function even without an outside air intake. That can't occur unless there is an air connection to the house.) Fans are available that install on the top of a chimney that can be speed controlled to ensure that air is moving up and out (see Figure 15.1). These systems can be used for single- family or multi-family buildings.

Then there is the damper in the flue. The chimney goes all the way up through the house and through the roof. It's a purposeful "stack." (They call them smoke stacks on ships.) Polluted air is supposed to move up through it

FIGURE 15.1   A chimney top fan. (Exhausto)

when there is a fire, but the air doesn't really care. It's like a teenager. It does what it wants. It will move up the chimney whenever the pressure conditions push up whether there is a fire burning or not. Closing the damper helps. If the damper is directly over the firebox, the chimney is full of cold air, and it's all sitting on top of that damper and that cold air will be oozing into the house whenever the pressure conditions pull down. Dampers are not very tight. The average effective leakage area of a closed fireplace damper is 30 square inches.

Dampers are available that mount on the top of the chimney. They close tightly, and they keep the air in the chimney warmer because it is full of house air as opposed being full of outside air. There are also inserts that can be installed in the fireplace when it is not in use that can be inflated to seal up the opening.

A source of intake air balances out the system when there is a fire burning. That air should be ducted from the outside. There should be a means for closing this "hole" in the house off when the fireplace or stove is not being used. Even though the opening in the house may be in the fireplace or stove, it will still be open to the house. If the house is under negative pressure, air will be sucked in through that pipe from the outside, air that will need to be conditioned.[1] A round exterior opening for a 4-inch duct is a 12 ½ square inch hole in the side of the house letting air in and out, 24 hours a day, seven days a week.

RADON VENTILATION SYSTEMS

Radon is a naturally occurring gas that can seep into homes and cause cancer. According to the Environmental Protection Agency (EPA), "Radon is the leading cause of lung cancer among non-smokers. It is the second leading cause of lung cancer in America and claims about 20,000 lives annually."[2] Homes should be tested for radon. It appears to be indiscriminate in terms of location. Just because none of the other homes in the neighborhood have a radon issue, it doesn't mean that your home doesn't. Testing should also take place over a period of time rather than a quick "snapshot." If the test is short term—only a day or so—the house may be under positive pressure and the radon gas will be kept out. There are numerous sources of information and products and services for testing for radon.[3]

The majority of radon gas enters the home from the ground (see Figure 15.2). The goal is to provide a means for the gas to by-pass the living areas of the house. Think of it like a lightening rod in reverse. Lightening strikes the rod and is conducted around the house through a wire to the ground where it dissipates. A radon mitigation system sucks the air out of the ground and ducts it up and out into the atmosphere.

If the house is under construction, there are radon resistant construction techniques that will help to air seal the foundation from a ground connection and that forms a sort of air chamber under the slab of the house that is connected to a pipe that runs all the way up through the house and through the roof. By "dumping" the polluted air out through the roof, it is less likely to flow back in through open windows or other inlets. An in-line fan that is capable of running continuously and strong enough to suck the air out from under the slab is added to the duct (see Figure 15.3). The system should also include a "fan-proving" indicator that clearly alerts the occupants if the fan system should fail. This is usually a pressure switch with a tap into the ducting that monitors the pressures on either side of the fan. If the pressures on both sides of the fan are equal, the fan is not running and a light or buzzer sounds.

1. www.hoyme.com for motorized dampers for 4-inch ducting.
2. www.epa.gov/radon.
3. www.radon.com.

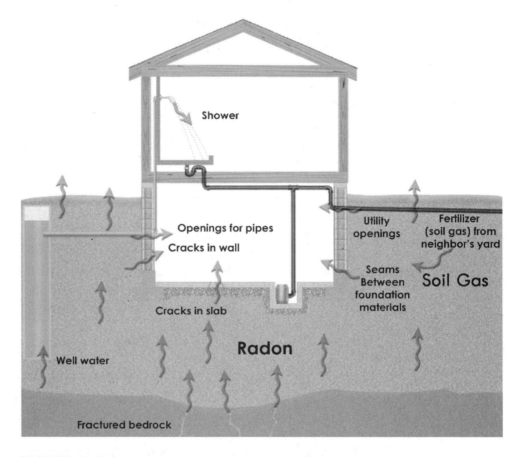

FIGURE 15.2 Radon sources. (Panasonic/Morrissey)

In an existing house, subslab depressurization can still be achieved. A channel can be formed around the perimeter of the basement if there is a gap between the slab and the basement walls. That channel can then be connected to the pipe that runs out through the roof. If there is a sump pump, the sump can be capped and air from the chamber can be tapped into the radon riser pipe. Some existing homes have drain tiles or perforated pipe that is designed to direct water away from the house. Suction from these tiles can be effective in reducing the radon gas that gets into the home.

If the house has a crawl space instead of a basement, the ground or floor of the crawl space can be carefully covered and sealed with a high-density plastic sheet. (This will also keep the humidity down in the space.) The air under the sheet can then be attached to the vent pipe and fan, providing "submembrane suction."

FIGURE 15.3   An in-line fan. (Fantech)

The piping and fan system need not be installed in and through the house. It can be outside of the house. This greatly simplifies finding a continuous path from the basement or crawl space all the way up through the roof. The fan that is selected must be suitable for installation in ambient conditions.

There is a great deal of discussion of the best way to handle the air in a crawl space. Many building scientists feel strongly that a crawl space should be sealed and insulated. But many building codes require that they be vented. All the issues of pipes and ducts and mechanical equipment in the crawl space must be considered for either approach. Power venting the crawl space to the outside for radon purposes can effectively remove the radon gas before it migrates up into the house. However, putting the crawl space under negative pressure may suck more radon gas into the space. Balancing the increase in gas with a flow rate to the outside is difficult at best. If there is mechanical

Different states have different radon construction codes. Sometimes information is available through the local department of public health. There is a great deal of information about radon issues at the EPA Website: www.epa.gov/radon.

equipment and ductwork in the crawl space, it is advisable to carefully seal the space up, and use a subslab or sub-membrane system to vent all the way out through the roof.

A positive pressure ventilation approach in the house may protect it from radon gas infiltration from the crawl space or basement. But as is true with all ventilation approaches, it is best to try to effectively remove the pollutant at its source like the moisture in a bathroom or the cooking smoke over a stove. There are so many variations in pressure in and on the house that it is virtually impossible to guarantee that the living space will always be under adequate pressure to keep the radon gases where you want them to be.

If a radon mitigation system is installed, it must be clearly marked. Once again, this is going to be one of those systems that runs all the time with no operator involvement. The "out-of-sight, out-of-mind" syndrome applies. The homeowner is not going to be interacting with the system at all. Even if there is an alert system on the fan, it needs to be clearly marked as to what it is and what a warning means. If the alert light or buzzer is behind a stack of boxes full of old clothes in a corner of the basement, the homeowner will never know what it is or why it is buzzing (if they can hear it). If they don't know why it is important to do something about it, they will probably just defeat it entirely.

Ideally, maintenance on the radon system should be on the list of other regular maintenance procedures. There is no filter to replace. It essentially consists of making sure that the fan is still running and that the alert system is connected.

Installed cost and operating cost will vary by the location, the system, and the house. The electrical cost of running the fan continuously is small. Most of the in-line fans consume less than 50 watts, many less than 20 watts. Some systems will be taking air from the house, which will need to be replaced with conditioned air, but the flow rates are extremely small. Total operating costs of $50 to $200 per year is pretty inconsequential if it reduces the risk of cancer.

The EPA site refers to using an HRV as an approach to radon reduction. All of the ventilation approaches discussed in this book including HRVs and ERVs will dilute the pollutants in the home, including radon gases. But the primary ventilation systems for improving the general air quality in the house are not directly coupled to the radon gas source. As such, an HRV or ERV would remove the smoke from the hearth fire if the chimney was eliminated, but it wouldn't be as effective. Radon is a serious health risk and radon gas should be vented directly with an independent system.

CRAWL SPACE VENTILATION

Codes require crawl spaces to be ventilated under most conditions. It would seem illogical to bring warm, humid air into a cold, dark crawl space and not expect there to be condensation and mold problems. Techniques have been developed and approved for sealing crawl spaces, making them a part of the living space, serving as a very short "basement." Many homes have their ducting and other mechanical equipment in the crawl, which requires careful attention to combustion air.

Codes do allow crawl spaces to employ mechanical ventilation in place of simple passive vents. Replacing passive crawl space vents with mechanical ventilation that draws air from the house and vents the crawl space air to the outside can reduce moisture problems and increase energy efficiency, but attention must be paid to the design of the system.

The 2009 International Residential Code § R408.3 (see Chapter 12) specifically allows crawl space design with mechanical exhaust. To comply the crawl space must have a continuously sealed, vapor retarding ground cover, have no passive openings to the outside, and employ a continuously operating exhaust fan. For the code minimum, mechanical ventilation must be provided at a rate of 1.0 cfm per 50 square feet of under floor space. The code does not specify what sort of fans should be used, but Colorado amended the IRC to require a fan rated for 44,000 hours (five years) of continuous operation with flex connections to reduce any fan vibration and resulting noise. There must also be some sort of indicator to alert the occupants if the fan should fail.

The system must also include transfer air openings, one per 250 square feet of crawl space floor area that are installed in the decking between the crawl space and the conditioned rooms above. To meet the Colorado code (and a

TABLE 15.1 Sizing crawl space transfer air openings.

Amount of Air Required	Minimum Hole Size (square inches)	Maximum Hole Size (square inches)
0–10 cfm	1.5	2.4
11–15 cfm	2.4	3.6
16–20 cfm	3.6	4.4

practical guide for other locations), the transfer air openings should comply with Table 15.1.

The crawl space exhaust fan airflow (typically 30 to 60 cfm) would be divided by the required number of openings to determine the flow per hole. The fan will exert a slight negative pressure on the house above. If there is an atmospherically vented appliance in the crawl space, the openings should be large enough so that the appliance will never experience a negative pressure condition of more than 2 Pascals (and the installation MUST be in accordance with all local codes for atmospherically vented appliances).

PASSIVE VENTILATION

Passive ventilation is ventilation without the benefit of a mechanical device like a fan. Opening the windows is passive ventilation when the air moves through them. Using windows for ventilation requires the right ambient conditions on both sides of the window, creating the pressure differentials that will force the air through the opening.

Passive inlets work with exhaust-only or supply-only ventilation systems to balance them out. Generally these are simply unplanned cracks and air leaks in the envelope of the house. There are passive inlet devices that have been dubbed "smart holes" that purposefully let air in more predictably. These devices were described in Chapter 8.

Natural or passive ventilation strategies have been around since they decided to put a hole in the roof of the house to let the smoke out. It was logical to take advantage of the stack effect or cross ventilation to get the pollutants out of the building. Central courtyards with fountains, so-called airshafts between city buildings, "solar chimneys," and building shapes have all taken advantage of prevailing natural conditions to passively compel air to move through buildings. To get passive or natural ventilation to work, it must be incorporated into the design of the building—the true house as a system approach. From that standpoint, considering a natural ventilation approach will enhance the overall building design and prevent it from working against itself. It is more difficult to design aggressive natural ventilation strategies into a traditionally designed home than it is to add a glass-enclosed stairwell, solar chimneys, or ventilation stacks into a commercial or industrial building.

There can be a thermal penalty for adding natural ventilation to a residence. Adding a space like a sunroom on the south side of the house can provide an "engine" to drive air circulation, but it will add heat at the same time. Strong temperature differentials can generate strong airflows, but it is not just for fresh air ventilation.

Early passive ventilation systems relied heavily on heat to provide the necessary buoyancy forces necessary to drive the airflow.[4] Natural or stack ventilation systems are extremely common in the United Kingdom, Northern European countries and Canada despite the fact that U.S. homes are built to generally lower tightness levels than homes in the other areas except for the U.K. This unintentional passive ventilation is known more commonly as infiltration and exfiltration. The potential advantage of passive ventilation is that using natural forces could avoid the energy that is required for mechanical ventilation. The challenge is achieving an adequate level of control in all seasons while avoiding the multitude of issues of dust and moisture, occupant intervention, potential drafts, and loss of conditioned air.

The European Building Research Establishment (BRE) has assumed the lead role in developing and promoting passive stack ventilation (PSV) systems. Their best practices system is concerned (as we are) with removing the pollutants at their primary sources in the home—bathrooms and kitchens. The systems consist of fresh air inlets and an exhaust stack that runs up through the roof with circulation openings in doorways or interior walls (Chapter 8).

According to the Axley article,[5] the BRE best practices PSV system includes:

- A service room inlet vent with an free area opening of 6.2 square inches (4000 square mm), intended to provide a low, continuous flow of inlet air, essentially a "trickle" ventilator;
- An outlet ventilator placed in the ceiling plane or high on the wall above the primary moisture source having a free area equal to or greater than the stack cross sectional area;

4. Axley, "Residential Passive Ventilation Systems: Evaluation and Design," AIVC, 2001.
5. Axley, p. 8.

- A ventilation stack duct, 5 inches (125 mm) in diameter for kitchens and 4 inches (100 mm) for bathrooms that conducts the air from the space all the way up to above the roof line;
- A stack terminal fitting that allows the air to flow out, maintains suction pressure, and prevents the infiltration of rain, insects, animals, leaves, and other debris.

Control and adjustment of the system have been traditionally left to the occupant through the use of adjustable louvers, but because the driving forces are in constant flux, this requires frequent if not continuous intervention and awareness on the part of the house's occupants. Passive inlet devices are available that automatically respond to pressure, humidity, or both, restricting the amount of incoming air to a prescribed level. These can make the PSV system far less dependent on occupant intervention.

On average these systems have been shown to work relatively effectively. Moments to moment ventilation requirements, however, are not well served. A sudden smoky event in a kitchen will be more quickly removed through the use of a local, well-designed and installed kitchen exhaust fan. People don't live in homes on an "average" basis. They live there all year 'round, in all sorts of conditions. But thinking again on a whole systems basis, a carefully designed combination of passive and active ventilation systems, a hybrid system, may satisfy both average and immediate condition. PSV systems can be designed and installed to meet the constant, low flow, background ventilation requirements of ASHRAE 62.2, despite the fact that PSV rates will never be truly constant as wind and changing house pressures drive them. They demand effectively addressing air circulation throughout the building, which has been shown to be a key element in removing pollutants. And they are particularly suited to tighter homes, as they provide controlled rather than random "leakage."

North American building codes have no specific provisions for passive ventilation systems beyond minimal window opening regulations for rooms that are not required to be mechanically ventilated. Because passive ventilation systems rely on pressures to function correctly, they will interact with other pressure reliant systems like fireplaces and gas water heaters. Implementation of passive ventilation strategies needs to consider possible interactions including code provisions that define the location on the lot, light and ventilation, dwelling unit separation, protection against radon, wall constructions, chimneys and fireplaces, mechanical exhaust systems, combustion air for chimneys and vents, and plumbing vents and traps.

Passive ventilation systems could be a useful tool in whole house systems integration, a means to cut one more piece out of the home's energy load. But without careful forethought and design, they are "delicate" systems in that they must be more carefully balanced than a mechanical ventilation approach. For example, when a thermostat turns on a heating system, the temperature and the amount of air delivered to the house is a known entity (all the maintenance issues set aside). As far as the thermostat is concerned, the air delivered to the house will warm the house. Working with an alternative source of energy, such as solar heat, the amount of energy is variable—it might be hot, it might be warm, or it might be cool. The driving forces in a passive ventilation system depend on the weather and on the occupants, variables that make it difficult to maintain control of the flow.

GARAGE VENTILATION

Attached garages are effectively just another room on the house where we put all the stuff that we don't want in the house. It seems that it is outside the living space, but unless careful steps are taken, what's in the air in the garage will be in the air in the house. There isn't much difference between the average garage and those big, glass doors that they open in automotive showrooms to drive the new cars in. Garages are located beside and underneath homes. There are rooms over them and, quite often these days, rooms under them. As houses have gotten bigger, garages have gotten bigger. They have room for more cars and sometimes boats. Mechanical equipment is located there—furnaces, boilers, air handlers. Ducting runs through them with all their associated gaps and leaks and pressure issues.

If the garage is not attached to the house, ventilation issues center around maintaining conditions in the garage for any activities that are pursued there. If, for example, there is a workshop in a detached garage, keeping it cool and keeping the air fresh by venting the fumes from a hobby or from the cars is a good idea. Any cooling system should be vented directly to the outside and not into an attic area and it should not overpower any combustion device in the garage. A means should be provided for supply air. The same is true for any "spot" venting system to remove the fumes over a workbench, for example.

To work on vehicles in the garage, careful consideration of venting exhaust fumes and providing make-up air is important for the health of the garage occupants. These measures may be as simple as an exhaust fan or as complex as a commercial exhaust fume ventilation system. As in other buildings, consideration of the interaction of other ventilation/exhaust systems should be kept in mind. If there is a dust collector, for example, it will be drawing out the air in the garage, working against a garage exhaust fan.

Note that the 2009 International Mechanical Code (IMC) includes a provision in Table 403.3 for 100 cfm per car in residential applications. "Mechanical exhaust required and the recirculation of air from such spaces is prohibited." IMC Table 403.3 page 33

Attached garages are far more of a problem because they are . . . attached. Building components join the garage to the living spaces in the house. Wall and ceiling cavities conduct air just as they do in other parts of the house. Moving air carries pollutants, and the garage is a haven for pollutants that are not good for human health. Ninety-three percent of the one million houses built in the United States in 2003 had one-, two-, or three-car garages.[6] Multiple studies and tests have been done in both the U.S. and Canada to determine the extent of carbon monoxide and benzene transfer into the house and determined that the pattern of carbon monoxide in the house mimicked the pattern in the garage.

Responding to many CO alarms, fire departments often have a difficult time determining the source of the CO and blame faulty detectors or the furnace because it's the most obvious combustion source. CO infiltrating from the garage is a slow process, commonly taking an hour or more with a low pressure differential between the two spaces. An elevated level in the garage can continue for four to six hours after the car has been removed. (Note that UL 2034 approved CO alarms are not permitted to alarm if the CO level exceeds 70 ppm for an hour but must alarm at that level before four hours have passed.) When a car is started in the garage, the engine is cold and it produces a huge amount of carbon monoxide, as much as 80,000 parts per million (ppm). Even if the garage door is open, the garage can fill with deadly CO within two minutes. The car should not be allowed to run for any length of time in the garage even with the door open.

Go back to the pressure discussions in Chapter 7. A hole located at the lowest point in the house will allow the most air in. A hole located near the top of the house will let the most air out. (This naturally depends on wind loading, etc.) A garage door is a very large hole and it is always located close to the ground. Unfortunately it is difficult to locate the garage entrance near the middle of the building! Homes and garages that are generally leaky have less of a problem with CO because of their elevated air change rates. (They

6. U.S. Bureau of the Census.

also have a higher energy cost.) CO is slightly lighter than air and moves easily throughout the house but it will not attach itself to surfaces in the garage and a complete air change will remove all the CO.[7]

Mild CO poisoning feels like the flu. More serious CO poisoning can lead to breathing difficulties, lack of reasoning, and death. It is likely to have the most serious effects on the young, the sick, and the elderly.

Before adding any sort of mechanical exhaust to an attached garage, the building system should be carefully analyzed. If the pressure in the garage is higher than the pressure in the house, air will flow toward the house carrying all the garage pollutants with it. Different configurations will require different ventilation approaches.

- What parts (if any) of the mechanical systems for the house are located in the garage? Is the air handler in the garage? Are there duct runs beyond the basic connections to the air handler?
- Is there a gas fired water heater in the garage? Is it a sealed-combustion system?
- Are there rooms above the garage? Below?
- Is there a stairway that leads to a basement door?
- How much of the house's infiltration comes from the garage? Sealing all the openings between the house and the garage will eliminate or significantly reduce that infiltration. (It is common for 25 percent of the total house leakage to come through the garage.)[8] That may have an impact on the ventilation and combustion appliances in the house.
- Consider the climate. (A garage in a cold climate may suffer from excessive condensation in the winter when a wet car is driven into the garage. In a hot climate, overheating may be a more important issue.)

The goal of any approach to improving the garage/house connection is to remove the garage from the house. This is difficult to do with an existing house, unless the garage can be converted to another room and a new garage built away from the house. This may not be an acceptable option because homeowners like the ability to walk from the garage to the house without going out in the inclement weather. So the next best step is to do everything possible to seal the connections between the house and garage (the house/garage interface) with the goal of disconnecting the air connection of the two but leaving them physically attached. Because some gases will move

7. For a free CO modeling tool, go to http://energytools.com/calc/garage.xls.
8. Wilber & Klossner, 1997 "A Study of Undiagnosed Carbon Monoxide Complaints."

TABLE 15.2 Effects of various CO levels.

Concentration of CO in Air	Effects
1–2 ppm	Might be normal, from cooking stoves, traffic, or outdoor air
2–8 ppm	Why is the CO level elevated? Source should be identified
9 ppm	Maximum allowable concentration for an eight hour period in any year (EPA, ASHRAE). Typical concentration after the operation of an unvented gas kitchen range. City air.
15–20 ppm	Impaired performance in time discrimination and shorted time to angina response.
20 ppm	Typical concentration in flue gases of a properly operating furnace or water heater[9]
30 ppm	UL standard requires that CO detectors not alarm at 30 ppm unless exposure is continuous for 30 days.[10] Earlier onset of exercise-induced angina
35 ppm	Maximum allowable outdoor concentration for one-hour period in any year, EPA, ASHRAE
50 ppm	Maximum allowable eight-hour workplace exposure (OSHA)
70 ppm	UL 2034 listed detectors must sound full alarm between one and four hours
150 ppm	UL listed detectors must sound full alarm between 10 and 50 minutes.
200 ppm	Maximum workplace exposure (NIOSH).
400 ppm	UL listed detectors must sound full alarm between four and 15 minutes.
500 ppm	Often produced in a garage when a cold car is started with the door open and run for two minutes
800 ppm	Dizziness, nausea, and convulsions within 45 minutes. Unconscious within two hours. Death within two or three hours. Maximum air-free concentration from gas kitchen ranges (ANSI)
1600 ppm	Headache, dizziness, and nausea within 20 minutes. Death within one hour. Smoldering wood fires, malfunctioning furnaces, water heaters, and kitchen ranges typically produce these concentrations.
3200 ppm	Headache, dizziness, and nausea within five to 20 minutes. Quickly impaired thinking. Death within 30 minutes. Concentration from a charcoal grille used indoors.
6400 ppm	Headache, dizziness, and nausea within one to two minutes. Thinking impaired before response possible. Death within 10-15 minutes.
12,800 ppm	Death within one to three minutes.
70,000 ppm	Typical tailpipe exhaust concentrations from cold gasoline engine during the first minute of cold weather start. (Concentrations decrease to 2 ppm after 17 minutes of running.)

Table adapted from Tom Greiner, Ph.D., P.E., Iowa State University.[11]

9. Greiner, Iowa State University.
10. UL Standard 2034, p. 35, Table 38.1.
11. Greiner, "The Silent Killer: Carbon Monoxide," Iowa State University, January, 1998.

by diffusion through some building materials (like block), it will be almost impossible to eliminate all of the transfer, but if it can be significantly reduced, the remainder can be vented more easily.

Every crack, gap, or hole between the house and garage must be sealed. Be merciless! Seal them up and get the house tested with a blower door. You could even get one of those theatrical smoke generators that they sell at Halloween and fill the garage with smoke and see where it goes. (Be sure to alert your neighbors first in case they think your house is on fire when smoke starts pouring out of soffits and the chimney and other unexpected places.) The primary leakage spots include:

- Basement headers;
- Pipe penetrations from the basement into the garage;
- Forced air heating supply duct chases in common walls;
- Plumbing penetrations in walls of laundry rooms or powder rooms next to the garage;
- A lowered ceiling abutting the house/garage common wall;
- Cold air returns passing through stud spaces in the common walls;
- Un-taped or damaged drywall joints on the garage side of the common walls;
- Pocket doors in the partition abutting house/garage common walls.[12]

It is not easy to find the leaks without a blower door or other specialized equipment, but the following symptoms may indicate that there is a problem that needs to be addressed:

- The carbon monoxide alarms go off frequently particularly when the car is started;
- Exhaust fumes are noticeable in the house and cause headaches or general, noxious smells;
- Cold drafts are felt in spaces adjacent to the house/garage common walls;
- The floors in the living space over the garage are commonly cold.

If there is mechanical equipment in the garage, make sure that it has adequate combustion air. Newer, more efficient furnaces and boilers send less waste products up their flues, creating a low stack pressure. It is ironic that more efficient combustion equipment is more susceptible to backdrafting sim-

---

12. Garage Performance Testing, The Sheltair Group, 2004, p. 26.

ply because of its efficiency. Even more care has to be taken in not overpowering these devices with competing exhaust equipment.

If the furnace fan is used for ventilation or continuous air circulation, sealing the house/garage interface may have little effect. The furnace fan is able to significantly reduce the pressure in the HVAC system to the extent that it will suck in and circulate any of the pollutants that are located in the garage.

Passive, through-wall vents between the garage and the outside will reduce the pollutants in the garage because they make the garage "leakier," but like other passive inlets and outlets, they don't always work in the desired direction and they can't be controlled to work when there is a pollutant level in the garage. If they are located high in the wall, they are more likely to serve as outlets than inlets. The passive vent approach works better if the house is tight and the house/garage common walls are sealed.

Finally, an exhaust fan can be an effective solution, bearing in mind the issues of combustion air, etc. The question is how big should the fan be? It partially depends on the tightness of the garage. Very tight garages can be effectively ventilated with very small fans (as low as 10 cfm) running almost continuously. In the majority of garages 50 cfm is about as small as the airflow can be to have any impact. Most local building codes are calling for 100 cfm per car. (Note that is a fully delivered, installed measurement, not a 100 cfm marking on the box.)

The size of the fan is also dependent on how it is controlled. If the purpose is to remove the high level of CO that is generated when a cold car is started in the garage, activating the fan when the garage door opens is a reasonable control strategy. The fan should continue to run for at least 30 minutes after the door closes.

A CO control can activate the fan whenever the carbon monoxide exceeds the set point whether the car is in the garage and running or has been removed and the CO is lingering in the air. It will also activate the exhaust fan if there is an elevated CO level from a barbecue grill, generator (as long as there's power!), or leaf blower. It's not always just CO that is a problem with the air in the garage, however.

A differential pressure control can sense the pressure difference between the house and the garage and activate the exhaust fan to keep the garage at a lower pressure than the house. In that condition, the pollutants can be kept out of the house, essentially flowing "downhill" from the garage to the outside.

Moisture is also a problem in a tight garage, particularly in a cold climate. Pulling a car or truck covered with snow into a garage that is above freezing will bring a great deal of moisture into the space. That moisture can cause the windows to fog and mold to grow.

Because of all these conditions, a continuously operating, low power fan in the garage is worth considering. If the fan is only drawing 20 watts, at $0.13 per kilowatt hour, it would cost less than $25 per year to operate the fan continuously. There isn't a conditioning penalty as there is for household air. This is another of those continuously running fans like the radon mitigation system that is definitely worth its operating cost.

DISPLACEMENT VENTILATION

Displacement ventilation is a strategy for introducing ventilation air near the floor ideally at a temperature that is lower than the air in the room and allowing it to rise to the ceiling where it is exhausted. The room is purposefully stratified, the warmer air "sitting" on the cooler, denser air providing cooler, fresh air into the area where people sit and work. Stale air polluted with carbon dioxide (CO_2) is pushed up to the ceiling.

This ventilation strategy can work very effectively in school classrooms or office spaces. Stale air is not recirculated. One hundred percent outdoor air can be supplied. Cooling capacity can be reduced because of the thermal stratification. Often less fan horsepower is needed and there is less room noise because the air velocity is so low.

Some applications for displacement cooling in residential applications have been tested and found to reduce cooling loads. Introducing conditioning air of any type near the floor in a house, however, runs the risk of having furniture put in front of it, blocking the flow. If the vents are in the floor, things cover them, stuff falls in them. The volume of air moving into and out of the room requires relatively large ducting, raising installation costs and requiring a sacrifice of space. There may be proven benefits in the future for residential applications if ventilation is combined with other conditioning functions.

Chapter 16
VENTILATION FOR COOLING

There was a time not that long ago that "ventilation" in the common vernacular meant a cool breeze wafting through the house, or simply a way to disperse the warm and stuffy air in the room (see Figure 16.1). Home occupants smell some pollutants, but they certainly feel overheated air. And while the nose gets acclimatized to smells, the body doesn't stop feeling the heat. More and more homes in all climates are installing central air conditioning. That technology has allowed us to move into areas that were previously considered uninhabitable for much of the year. Ventilation can be used to cool down the house and cool down the people. It can get rid of excess heat in the attic.

There is some controversy about ventilation that is used for cooling, particularly if we start at the top of the house. Is it necessary to remove the heat from the attic? Does it accomplish anything in terms of cooling load? Should the same rules for cooling the attic be applied in all climates? Attics are one of those simple spaces that have become complicated like garages and crawl spaces. Before tight building materials and high levels of insulation, the attic was just an empty space at the top of the house. If life got crowded, one of the kids moved up there. Ice dams and humidity are attic issues but they are not directly "cooling" issues, which is what this chapter is about. But when considering an attic, air motion for cooling is also air motion for removing moisture.

Attic ventilation has two distinct purposes: 1. Controlling the temperature of the roof primarily for ice dams and roofing material manufacturer requirements (the passive ridge and soffit ventilation system), and 2. Cooling the attic for mechanical equipment and stuff stored there and reducing the temperature of the blanket of air that sits at the top of the house (powered attic ventilators or attic fan or passive gable vent systems). These are two different systems with two distinct purposes.

RIDGE AND SOFFIT VENTS

Soffits are the underside of the overhangs at the bottom of the roof. The ridge is the top or peak of the roof. The object of the combination of these two vents is to allow air to enter at a low point and flow passively to the high point, cooling the roof, a reasonable concept. For it to work effectively, the path for entry and exit must be clear. If insulation or other attic detritus is jammed down on top of the soffit vents, the airflow path will be blocked. At the ridge, quite often the roof has not been cut open and there is no way for the air to leave.

It is difficult to know what the ridge or soffit vent is capable of doing as a product. How much air should be able to move through it if the conditions are right? How much of an opening does it provide? Product manufacturers calculate the area of the openings and arrive at what they call the net free area (NFA). That doesn't provide the whole picture. The configuration of the product will affect its flow resistance as it does with fans or grilles. The best way to determine the ability of the product to allow air to flow through it is to measure the flow of air through the product!

The Home Ventilating Institute (HVI) has a procedure for testing what they call "static vents." A test stand is set up that imitates how the product will be installed on the roof. Air is pushed or drawn through the product at four separate static pressures. The rate of flow is plugged into the equation:

$$NFA = (0.0592Q)/\sqrt{\Delta P}$$

where

NFA = the effective net free area

Q = the airflow in cfm

ΔP = the difference in pressure (in this case the static pressure measurement point at standard atmospheric density)

Manufacturers of HVI certified products like Quality Aluminum Products put their products through this procedure to provide an accurate indication of the effective net free area. This is what should be stamped on the product and used to determine the adequacy of meeting the 1:300 attic ventilation standards for attics with a vapor barrier separating them from the living space. Tests have been established by the International Conference of Building Officials (ICBO) and the University of Illinois as well as HVI to determine what they call an equivalent net free area or ENFA. But the protocols are different and the manufacturers aren't willing to promote the resulting lower performance number, so unfortunately consumers are stuck with what in many cases is erroneous performance data.

In terms of temperature change, what effect does the ridge/soffit venting approach have? There is about a 30°F difference in attic temperature between an attic with no vent cap—just shingled over—and an open ridge vent slot (not a real world, practical approach due to weather issues.)

Attic ventilation was added to building codes to prevent roof damage caused by moisture laden air migrating from the living space during the winter when that warm moist air hits cold surfaces, condenses, freezes, and causes mold problems. The change was implemented after the introduction of indoor plumbing and central heating systems. With the introduction of increased attic insulation ventilation proved valuable in preventing ice dams. If the entire roof deck can be kept at the same temperature, snow will melt off it uniformly and not cause a problem. But if the edges of the roof, hanging out

FIGURE 16.1   A room fan. (Air King)

beyond the attic area, are colder than the middle, snow melt will freeze there and form a dam, pushing the water up under the roof shingles. Keeping a flow of air moving up the underside of the roof sheathing will keep it at a uniform temperature.

As with many technologies, layered in tradition, eventually these original purposes were forgotten, particularly in cooling climates where ice dams are not an issue. The thought was that attic venting could lower the attic temperature, which would reduce the cost of cooling and prolong shingle life even though there was no research to support those concepts. Adding an active cooling system or fan was assumed to amplify the effect, but field testing has shown that the "temperature of roof sheathing of a [sic] unvented roof will rise by a few to no more than 10°F more than a well ventilated attic."[1] Despite this, shingle manufacturers generally insist on the vented deck design and most codes require it.

If the deck is to be ventilated, there should be a minimum 2 inch clearance between the bottom of the deck and any insulation or other air blocking materials. This will effectively allow for a 2 inch x 22.5 inch "duct" in each cavity (if the rafters are 24 inches on-center), "open" at the bottom and top, forming an effective solar chimney, passively venting each cavity. Baffles must be used to prevent "wind washing" of the insulation at the eaves, which reduces the thermal resistance of the insulation. Choose your soffit and ridge vents carefully, looking for HVI certification. Compare the actual openings in soffit vents. Visually you can often tell if one is more open than the other despite the claims. Neither may be accurate, but a visual inspection can provide logical clues. The more open it is, the easier it will be for the air to flow through it. So called "button-louvers" are almost impossibly restrictive to the airflow.

Some ridge vents include baffles to prevent the air from simply blowing through from one side to the other. Winds and breezes will drive the air up the outside of the roof and over the ridge, creating positive pressure on the side facing the wind (the windward side) and negative pressure on the opposite side (the leeward side). By adding the baffles more air may be forced over the top, creating more negative pressure and sucking more air out of the attic. Although this may have some effect on performance, the clear openings are the most important performance factor.

---

1. "Understanding Attic Ventilation," Joseph Lstiburek, Building Science Digest 102, 2006.

ATTIC FANS (POWERED ATTIC VENTILATORS)

If passive vents work so well, why not add some mechanical assistance and put fans in? Once again, a seemingly simple and logical ventilation approach is not as simple as it appears. There are a lot of articles on the Internet about why it is so important to use power attic ventilators (PAVs) or attic exhaust fans. Companies that make PAV products have produced most of these, but they can't all be discounted. Manufacturers get a lot of feedback about their products—both good and bad. How much they listen is another question.

There is no doubt that attics get hot. Ridge and soffit vents are not there to significantly reduce the temperature in the attic. If it is important to reduce the attic air temperature, something more is needed.

If the attic is completely sealed off from the house, then the problem is limited to the attic, and adding more airflow will affect only the attic. But in the majority of homes, the top floor ceiling plane is not well sealed. There are a myriad of holes from recessed lights, wiring boxes, wall stacks, plumbing stacks, chimneys, etc. Lowering the pressure in the attic with an attic exhaust fan will draw air from the house up into the attic. If the fan is powerful enough and the house is closed up, that could depressurize the house enough to back-draft combustion appliances or draw radon or soil gas in from the crawl space. This is just another of those "house-as-a-system" issues that you must think about. Air sealing the ceiling plane solves many problems in both warm and cool weather. Keeping the mechanical equipment out of the attic solves another pile of issues like servicing and limiting duct leakage to the outdoors.

Another air conflict may arise between the passive ridge/soffit system and a powered attic ventilator. The PAV will draws its make-up air from the closest place, which may be the ridge or soffit vent, short-circuiting the path up the underside of the roofing material. Adding any PAV requires adding an adequate, intentional source of make-up air. (Remember: one cfm in equals one cfm out.) HVI recommends 1 square foot for each 300 cfm of airflow (see Table 16.1). Remember that these are "clear" openings to the outside. Weather louvers will reduce the opening size because they block some of the opening area. Insect screens will severely reduce the openings so the overall hole size will need to be bigger than these numbers.

The tighter, better sealed the attic space relative to the outside, the larger the hole needs to be to balance the airflow exhausting through the PAV. PAVs are generally axial fans and don't work well against backpressure or flow resistance. Most of them will operate at about half the flow rate printed on the box once they have been installed because of the added resistances of grilles, louvers, and inlet openings.

TABLE 16.1 Exhaust fan inlet hole sizes.

Exhaust airflow cfm	Inlet area in square feet
300	1
600	2
1200	4
1650	5.5
3000	10
3900	13

In terms of the effect of the cost of cooling the air in the house, reducing the attic air temperature has only a small effect. Warm air flows upward. The hot air in the attic will not flow down into the house. In terms of conduction, if the floor of the attic is poorly insulated, the ceiling of the house will reach the same temperature as the attic, and that heat will be conducted and radiated into the house. You can feel it if you stand under a poorly insulated attic hatch. If the floor of the attic is well sealed and insulated, then the attic temperatures will be isolated from the house, and the large temperature swings up there are less important.

If the mechanical equipment and the ducting are in the attic, reducing the air temperature is much more important than if it is just empty space. The floor of the attic may be covered by R30 insulation, but the ducting is only wrapped in R4, R6, or at most R8. If the attic air temperature is 140°F and the house temperature is 72°F, that is a 68°F temperature difference across R30 insulation. The air temperature in the ducting may be 60°F, providing an 84°F temperature difference across R8 insulation. Ducting may cover 25 to 30 percent of the attic floor area, but the total surface area of the ducting exposed to the attic temperatures is much greater than that. Every five degrees of temperature drop will save 1.6 Btus per linear foot of R8, 10-inch diameter ducting, per hour the air conditioning is running (assuming that the entire perimeter is exposed to the attic air temperature). (See Table 16.2.)

TABLE 16.2 Attic temperature effect on duct insulation.

5°F reduction in temperature per linear foot per AC operating hour	12-inch Flex Duct	10-inch Flex Duct	8-inch Flex Duct
R8 insulation	2 Btus	1.6 Btus	1.3 Btus
R6 insulation	2.6 Btus	2.2 Btus	2.6 Btus
R4 insulation	3.9 Btus	3.3 Btus	2.6 Btus

So if there are 200 linear feet of 10-inch R6 ducting in the attic of an 1800 square foot house and the AC runs 30 percent of the time or eight hours per day: $200 \times 8 \times 2.2$ Btus = 3520 Btus per day per five degree temperature reduction. At an optimistic sensible EER for the AC system of 10 Btu/Wh, this would be a daily savings of 0.35 kWh. At \$0.129 per kWh the savings would be approximately \$0.045 per day. Adding that to the savings through the R30 insulation on the floor of the attic of another \$0.093 would indicate a savings of about \$0.14 per day for each five degrees the attic temperature was reduced. Although these are approximate numbers, they do provide a scale. As the attic temperature approaches the outside temperature, the ventilation system's affect will be reduced. It will take more and more air to remove the last few degrees.[2]

If the attic space is used for storage, it will be important to maintain a reasonably temperate environment. Moisture as well as temperature is a factor in maintaining the integrity of personal treasures. Moisture will come primarily from air leaks between the house and the attic (or the practice of ventilating bathrooms into the attic, or laying the bathroom ducting "near" a gable or soffit vent in hopes that it will find its way outside). Roof or chimney leaks are another potential source of moisture and obviously need to be sealed/repaired.

Passive vents offer an alternative to PAVs. Most of these are similar to very small dormers or "eyebrows" that are mounted on the roof. They are larger and far more open to airflow than ridge or soffit vents because they are cooling the temperature of the air in the attic, not just reducing the temperature of the roof deck. There are also "wind spinners" that are more commonly found in commercial buildings, the top spinning around in the response to breezes, establishing a vortex in the air inside their pipes that sucks the air out of the attic.

One company[3] has a passive vent cap that employs directional louvers to establish the vortex flow inside the pipe. These can be effective in low wind speeds to establish low exhaust flows that exceed passive stack effect ventilation rates.

There are a number of solar powered attic exhaust fans. The beauty of these is that the fan runs when the sun is shining and the attic is heating up. Because fans will bring in outside air that is cooler but can also be full of moisture, solar powered fans have the advantage of running only during the

2. Danny Parker of the Florida Solar Energy Center (FSEC) noted that "attic ventilation can make a difference but not nearly so great as that achieved from a light colored roof." See "Comparative Evaluation of the Impact of Roofing Systems on Residential Cooling Energy Demand in Florida" on the FSEC Website, www.fsec.ucf.edu/en.
3. Active Ventilation Products.

daylight hours when the relative humidity levels are generally lowest. Different sized photovoltaic (pv) arrays produce different amounts of power, producing a variety of airflows from the fans. The power of the pv array is directly related to the air movement capability of the fan.

WHOLE HOUSE COMFORT VENTILATORS (WHOLE HOUSE FANS)

A whole house comfort ventilator, or what is commonly referred to as a whole house fan, is a fan that moves the air from the living space for cooling. They are not used for improving the air quality in the house except for reducing the temperature by exchanging the existing, warm house air with cooler outside air. They have no mechanical cooling capacity beyond their air moving capabilities so if it is hotter outside than it is inside, they shouldn't be used.

The traditional design is a large, axial fan mounted in the top floor ceiling between the house and the attic. The ceiling side is covered by a grille and/or passive louver that is sucked up and open when the fan is turned on. They have been very large fans, 2 to 4 feet in diameter that can move thousands of cubic feet of air per minute, changing all the air in the house four to six times per hour. The point of these large airflows is to create breeze that moves through the house to cool the people. Unfortunately, fans that are strong enough to move that large a volume of air, are as a rule, big, noisy, and power hungry, probably not the sort of thing you want running all night long in your bedroom. To have the maximum effect on the largest volume of the house, they are generally mounted in a central location like in the hallway. Bedroom doors and windows need to be left open to allow the air to circulate properly, keeping the convective cooling going. Opening first floor windows allows the air to be drawn in and pulled through much of the house. The fan pushes the house air into the attic, reducing the temperature of the attic air and pressurizing the attic. If there are adequate relief vents from the attic, the air will be

Whole house fans are NOT the same as attic fans or PAVs even though a number of knowledgeable Web pages identify them that way. Attic fans (as discussed previously) are for reducing the temperature of the air in the attic. Whole house fans or whole house comfort ventilators are for reducing the temperature of the air in the house.

Whole house comfort ventilators work by depressurizing the house. They will draw air from any opening, always taking the biggest, easiest path. They are designed to be powerful and move a lot of air. NEVER operate a whole house comfort ventilator with the windows closed. The fan will almost certainly suck the air down a chimney or a combustion appliance. The switch for the fan should be clearly labeled. If an automatic control like a timer or thermostat is used, be sure windows are open before leaving the fan unattended.

forced to the outside. If the attic relief vents are too small, air will find its way through any opening, sometimes into wall cavities, pushing attic air and dust back down into the house, creating that sort of "burnt" stuff odor. At the same time the effectiveness of the fan will be reduced because it will be working against a higher pressure and its airflow will be reduced.

As the air in the house warms up during the course of the day, all the materials in the house warm up, absorbing and holding heat. (Heat moves toward cold.) Circulating cooling air through the house allows the heat flow from the materials to reverse, and the materials give up the stored heat to the cooler air, warming the air. The greater the temperature differential, the more quickly the materials give up the heat. "Eat your supper before the food gets cold!" Moving more air across the materials will cause it to give up its heat more quickly (which is why people blow on hot soup before they put it in their mouth). But air is not all that good at holding heat. A kitchen match produces about 1 Btu (British thermal unit) of heat. One cubic foot of air can only hold about .018 Btus for each degree of temperature change. So if it is only one degree cooler outside than inside the house, you need a lot more air to bring the temperature down than if it is 30 degrees cooler. Turn on the whole house fan when it is 40°F outside and you'll cool the house down pretty quickly! But if it is 70°F inside and 69°F outside, it will take a lot of air to cool the mass of the house.

You can use a really big fan and move the air quickly or you can use a smaller fan running over a longer period of time. It is common for the outside air temperature to drop over the course of the evening and night, and a smaller, quieter fan can take advantage of the increasing thermal effect.

As houses become more energy efficient, smaller fans work even more effectively. Because of higher insulation levels, the temperature inside the house will not rise as much in warm weather as less well insulated homes. A smaller fan working over a longer period of time can produce a gentle flow of air through the house, cooling the mass and making the occupants comfortable.

When thinking about the installation of a whole house comfort ventilator, think about the complete system and how you will be using the product. Because these products rely on outside air for cooling, windows will need to be open when the fan is running. That does mean that whatever is in the outside air will be pulled into the house. Window screens will be the primary filtration system.

You will also want to carefully consider the exhaust side of the system. If the fan is venting into the attic, the attic pressure will be raised. Providing adequate exhaust openings from the attic is critical to the performance. Dividing the listed airflow of the fan by 765 will provide a good approximation of an opening from the attic space to the outside to minimize the backpressure on the fan. For example, if the whole house fan is rated at 4000 cfm, the exhaust opening: $4000/765 = 5.23$ square feet. Again, note that this is "free area" opening. Insect screening will cut the open area by 20 to 30 percent. Louvers may cut another 20 percent. Since whole house comfort ventilators will only be used in hot weather when ice dams are not a concern, including the ridge and soffit vents in the opening calculation is acceptable.

How quickly a whole house comfort ventilator reduces the temperature in the space depends on the size of the house, the temperature of the "mass" of the house, and the temperature difference between the inside air and the outside air. As the temperature in the house approaches the outside temperature, it will take longer and longer to cool those last few degrees. It's the old question of, "How many steps will it take to walk to a wall if you walk half the distance each step?" Each step becomes smaller and smaller.

As long as it's cooler outside than inside, a whole house comfort ventilator can provide essentially free cooling. For example if it is 65°F outside and 75°F inside for an hour, a temperature difference of 10°F, and the fan moves 1800 cfm: $1200 \times 10 \times 1.08 = 19,440$ Btus, more than a ton of cooling (1 ton of cooling equals 12,000 Btus). Small central air conditioners deliver 2 tons of cooling. The cooling ability of a whole house comfort ventilator is not as predictable as an air conditioner, of course, but a ton or more of free cooling is not something to scoff at.

In terms of airflow, dividing the volume of the house by the cfm capacity of the fan provides the number of minutes per air change. A 2000 square foot house with 8 foot ceilings would have a volume of $2000 \times 8 = 16,000$ cubic feet. A 4000 cfm fan would move all the air in four minutes. An 1800 cfm fan would move all the air in nine minutes.

When the weather cools and the whole house comfort ventilator isn't needed anymore or when the weather is really hot and the central air conditioning is running, the opening for the fan needs to be closed. The gravity louvers on a typical whole house fan have a lot of "edges." Each one of those edges is a potential air leak. Stack pressure in the house will constantly push up on the louvers, and there will be a continuous trickle of air leakage. Providing a winter cover for the fan is an important seasonal step. Any control system needs to be defeated or carefully marked so that the system can't be activated with the winter cover in place, otherwise the fan will be burned out.

Some smaller, whole house comfort ventilators have motorized and insulated covers that automatically close when the fan is turned off (see Figure 16.2). These will provide a positive air tight and insulated seal. Make sure the edges of the fan are insulated as well as possible, adding loose fill insulation around the perimeter. Some of these small units blow the air out of the sides, horizontally. Loose fill won't help those units. They will need to be installed high enough above the insulation.

There are several ducted versions of whole house comfort ventilators. Some of these use large, 14-inch diameter flexible ducting attached to a panel fan located in the attic. As described in Chapter 14, axial fans are not good at

FIGURE 16.2 Whole house comfort ventilator set in place. (PHR).

moving air against any kind of resistance. Flexible ducting has a high resistance, which will degrade the performance of the fan. Be sure that the complete whole house comfort ventilator system—including the grille, the ducting, the fan, and the backdraft louvers—has been certified for performance by a third party like HVI. Otherwise the manufacturer may be just citing the "free air" (no resistance) performance of the panel fan. One nice thing about these products is that the fan is located remotely from the living space and they can be very quiet.

Another version of this product is a multi-port approach using a motorized impeller that is designed to pull air through flexible ducting (Figure 16.3). It is also designed to vent directly to the outside so that it is not relying on attic openings and will not interfere with ridge and soffit ventilation. With multiple ports, it can draw air from multiple, individual rooms, meaning that bedroom doors can be closed at night, allowing for privacy and maximizing the cooling performance in the bedrooms where people are sleeping. The bedroom window can be opened a few inches, making the total volume of air to move through a smaller opening, greatly increasing the velocity of the airflow.

Some of the issues in deciding on the use of a whole house comfort ventilator are highlighted in Table 16.3.

TABLE 16.3   Whole house comfort ventilator variables.

Pros	Cons
Enhances "natural" cooling by accelerating the transfer of cooler outside air through the house to the outside	Brings in large amounts of outside air and all the things that are in it including dust, pollen, and humidity
Less expensive to operate than air-conditioning systems	Depressurizes the house and may interact with other air sourced appliances like water heaters
Uses natural, outside air (not everyone likes air-conditioning)	Uses natural, outside air
Can be quiet enough to provide a background white noise that can mask outside traffic noises	Can be noisy (minimized by using smaller or remotely mounted fans)
Big fans can provide a rapid dump of over-heated air to reduce the air-conditioning load	Big fans are generally not well sealed and heat from the attic will radiate down through them
House air is considerably cooler than attic air so pushing attic air out means the whole house comfort ventilator is serving two purposes	Should never be used when the outside air is hotter than the inside air, and if the fan vents into the attic, the pressure can force attic air back down into the house.

FIGURE 16.3   A ducted whole house comfort ventilator. (R.E. Williams)

In selecting a whole house comfort ventilator be sure that the fan is safety tested by Underwriter Laboratories (UL), Intertek (ETL), the Canadian Standards Association (CSA), or another recognized independent safety laboratory. If you want to be sure that the airflow the manufacturer is claiming is real, it also needs to have been performance tested by a third party laboratory like HVI or AMCA. Sound level will be an issue with the larger fans. Some of them offer multiple speeds that will allow some flexibility in how the fan is used. Even some of the smaller fans are noisy and generally don't offer speed control capabilities.

Sizing, determining the cfm rate of the fan, depends on what you want the fan to do as described in Table 16.4. The traditional approach was to calculate the volume of the house and then multiply that by ½ or one. So if the house has a total of 1500 square feet of floor area with 8-foot high ceilings: 1500 × 8 = 12,000 cubic feet. Half of that would be 6000 cubic feet per minute. That fan size would change all the air every minute and would require almost an 8 square foot free area opening from the attic! It would be difficult to store anything in the attic with that much air moving through it. One builder

TABLE 16.4   Whole house comfort ventilator decision tree.

If you want to . . .	Consider . . .
Change all the air quickly, creating a breeze and "flushing" the house out before turning on the air-conditioning . . .	A large, traditional design whole house fan, something that will change the air every four or five minutes.
Cool the mass of the house over an extended period of time and push the air out through the attic vents . . .	A smaller, centrally located fan, or a number of smaller fans located in individual rooms.
Cool the mass of the house over an extended period of time, push the air out through the attic vents, and sound level is a major concern . . .	A ducted, attic mounted fan with a central or multiple vents.
Cool the mass of the house over an extended period of time, sound level is a major concern, vent from a number of rooms, and not pressurize the attic . . .	A multi-port attic mounted system, with a direct vent to the outside.

expressed the opinion that fans that big put so much pressure on the attic space that it will push the nails out of the roof! That may have been an exaggeration, but that is the sizing method the government flyer on whole house fans suggests and is repeated in many informational pieces. As described previously, selecting a fan with a lower airflow rate just takes a bit longer and allows the house to cool down more naturally.

PEDESTAL AND BOX FANS

Smaller, whole house comfort ventilators do a great job of cooling down the mass of the house. Pedestal fans, desk fans, or hassock fans provide portable means of cooling people (see Figure 16.4). They can be set up in any room. They are almost always capable of multiple speeds and many of them oscillate, turning back and forth, blowing air throughout the room. They are made

Not all fans have been tested for use with solid-state speed control devices. In fact, some installation instructions will specifically state "Not for use with solid state speed controls" if they have not been safety tested. They can also create motor hum and may burn out the motor.

FIGURE 16.4   A pedestal fan. (Air King)

in many sizes and styles, from really inexpensive units you can find in drug stores to solid industrial types that are used in factories, known as "man coolers." They all have the common characteristic of being simply designed to circulate air in the room without any ducting resistance, sucking the air in one side and blowing it out the other.

Unless it is an industrial or commercial application where the fan will be subjected to fairly polluted air and run almost continuously, these products are remarkably durable. Most of the variations are cosmetic, and there are certainly variations in the quality of the materials and the assembly. The grilles may come loose and the base may become wobbly, but for the most part they seem to last a remarkably long period of time.

Pedestal or oscillating fans move air around inside a room. They do not change the temperature of the air. Leaving a pedestal fan running when the room is unoccupied will not change the room's temperature unless there are other sources of conditioning like a room air conditioner or space heater.

Most of these fans have height adjustments. Be sure that the range adjustment works for your application. You may not want it blowing on you all the time. Some fans also have an angle adjustment, allowing you to tip the fan back so the airflow sweeps across the ceiling, again limiting the direct draft.

The weighted base versions help to keep the fan from getting knocked over easily, but also making them less portable. If you want to put your fans away up in the attic in the fall when the cooling season is over, you want a version with a lighter, un-weighted base. Some products come with wheels on the base, but that probably means that the fan is fairly heavy. The wheels will help to maneuver the fan around a limited area.

The more industrial grade versions can be noisy. In a factory setting, the sound level isn't all that important. In your living room where you are trying to have conversation or watch TV, noisy fans are unacceptable. Having multiple speeds can generally quiet most of the fans down to a reasonable level, but it is worth reading some reviews before making a purchase. The speed controls that have a limited number of settings generally mean that the fan has a number of "windings" on the motor, providing fixed points for the speed selector to connect to. That means that whatever setting you select the motor will operate normally. If the speed settings are infinite, then the controller will be reducing the energy supplied to the fan. The fan may hum and its life may be reduced. Having a handful of settings is probably enough to satisfy most people.

The descriptions on these products do not always tell you how much air the fan moves. This is probably because most people don't know what cfm means, and whether it is 750 cfm or 1000 cfm doesn't really matter. The fans are commonly sized by the diameter of the blade: 16 inch, 20 inch, 22 inch, 30 inch. Larger blade diameters will move more air. For most household applications, the smaller diameter fans are more than adequate. A 30-inch fan may move as much as 10,000 cfm at high speed, a 24-inch may move 8200 cfm, and a 12-inch fan may move 1000 to 1200. That's still enough air motion to push the pages of magazines around on your coffee table. Generally what is needed in a residential application is enough air motion to provide a gentle, barely noticeable, continuous breeze.

Box fans are generally used as window fans. Commonly they are square, box-like, with a handle on the top to make them portable. Installed in a top floor window, they can serve as a portable whole house comfort ventilator. They should be adjusted so that they are blowing out of the house when they are used on the second floor. That way they will be drawing air up from the first floor and cooling the whole house. For the best effect, the room door should be left open, allowing the air to flow both into and out of the room. (1 cfm in equals 1 cfm out!)

Some of these fans are less "boxy" and may actually be two fans, side-by-side, but whatever the design, they are more often used as space coolers or space ventilators than "man coolers," like the pedestal fans.

CEILING FANS

There are so many variations of ceiling fans on the market that it is almost difficult to describe them. There are wooden ones, plastic ones, stainless steel ones, fans with lights, fans with heaters, fans that clean the air, but the basic, common purpose is air movement—slow, gentle, pushing the air around in the room (see Figure 16.5).

They are available in a wide range of costs. The motor grade is a major cost factor. The least expensive ceiling fans are not designed to run for more than eight hours per day. All motors have a life expectancy, so you should figure on replacing the fan at some point in the future depending on how much it is used. Some motors have sealed bearings that require no maintenance. Others will require periodic lubrication.

Most ceiling fans are relatively low power—in the range of 50 to 75 watts. Some have aerodynamically optimized blade design to minimize power consumption and maximize performance.

All Energy Star rated ceiling fans include the use of a control that allows the ceiling fan to be used throughout the year. In hot weather, the fan should rotate counter-clockwise, pushing the air down and causing the occupant to feel a gentle breeze. In cool weather, the fan should operate clockwise at a low speed, creating a gentle upward flow, accelerating the natural upward movement of warm air. The fan causes the airflow to impact the ceiling, sweeping the air back down into the room. (Note that ceiling fans have no mechanical cooling component. Leaving them running in an unoccupied room will not cool the room.)

The Environmental Protection Agency Energy Star program makes the size recommendations shown in Table 16.5.

FIGURE 16.5   A ceiling mounted fan. (Broan)

Ceiling fans provide an effective means to make the room occupants more comfortable by moving the air. In the cooling season, raising the thermostat by just one degree can reduce the air-conditioning bill by about 4 percent over the course of the cooling season. Lowering the set point by one degree in winter results in about 3 percent savings in overall heating cost. Many energy efficiency programs give credit for having ceiling fans for just this reason. When air isn't moving, the room can feel stuffy. However, if the air is too cold, moving it around can make it feel drafty, at which point the ceiling fan should be turned off. It is not one of those automatic products. It requires operator involvement to be effective and reduce energy costs.[5]

COOLING TUBES

Earth tubes or cooling tubes are an unusual cooling/ventilation strategy that was explored in the late 1970s that seems to be of interest once again. The concept is that air is drawn through pipes that are run through the ground,

5. Sonne, J. and Parker, D., 1998, "Measured Ceiling Fan Performance and Usage Patterns: Implications for Efficiency and Comfort Improvement," Presented at the 1998 ACEEE Summer Study on Energy Efficiency in Buildings, Vol. 1, pp. 335-341.

TABLE 16.5   Energy Star ceiling fan sizing.[6]

Room Dimensions	Suggested Fan Size
Up to 75 square feet	29″–36″
76 to 144 square feet	36″–42″
144 to 225 square feet	44″–50″
225 to 400 square feet	50″–54″

6. http://www.energystar.gov/index.cfm?c=ceiling_fans.pr_ceiling_fans_basics.

6 to 12 feet below the surface. Since the ground temperature can be relatively stable and cooler than the above ground temperatures, the air is "conditioned" as it moves through the pipe into the house.

Experiments were done at the Passive Solar Research Test Facility at the Omaha campus of the University of Nebraska in 1982 that produced a significant volume of data on nine tubes of different sizes and lengths and buried at different angles. They confirmed that the maximum cooling effect for any tube is "heavily dependent upon the difference between the ambient outdoor temperature and the soil temperature surrounding the tube."[7] They also determined that depth of the tube and night ambient conditions played a role in maximizing the cooling effect.

As the air passes through the tubes over the course of the cooling season, the surrounding ground temperature will rise as it absorbs the heat from the air moving through the pipes. The diameter of the pipes, the number of pipes, and the length of the runs all have an impact on the effectiveness of this strategy. Lower airflow velocity at the mouth of the inlet allows the slower moving air to give up more heat to the ground. Sophisticated control strategies were developed that varied the fan speed with the changing temperature.

There are numerous reasons why the strategy is not more commonly used today. Running warm, humid air through pipes underground will naturally cause condensation to occur on the walls of the pipe. Moist surfaces in dark spaces will grow mold. Insects, animals, and soil gases can all enter the pipe in both the short term and the long term. And installing long length and circuitous paths of pipe in the ground can be an expensive process.

Yet the experiments continue.

7. Chen, Bing; Wang, Tseng-Chan; Maloney, John; Ennenga, James; and Newman, Mark; "Measured Cooling Performance of Earth Contact Cooling Tubes," Passive Solar Research Group, University of Nebraska, 1982.

EVAPORATIVE COOLERS

This is a topic that should fall under a space-conditioning text, but the evaporative cooling process relies on pressurizing the house, pushing a lot of air in and allowing it to "leak" out. They are sort of like whole house comfort ventilators in reverse with a mechanical cooling component.

Evaporative coolers, or "swamp coolers," rely on the fact that as water evaporates it removes heat from the surrounding air. They rely on low ambient relative humidity, dry air that can absorb a significant amount of moisture. A large volume of air is pushed or pulled through a wet medium and as the water evaporates the air is cooled. The cooler air flows into the house and is allowed to escape through open windows or openings into the attic. They use less energy than most mechanical cooling systems. The pads, which are often made of wood shavings, from wood like aspen that resists mold growth, have to be replaced every season or two.

The systems can work effectively at fairly high ambient temperatures as long as the relative humidity is low. When the relative humidity (RH) is as low as 2 percent, an evaporative cooler can drop 105°F air to 72°F. But if the RH rises to 40 percent, the 105°F air can only drop to 89°F. An advantage evaporative coolers have over whole house comfort ventilators is that they work best during the hottest time of the day because the RH drops as the temperature rises.

Evaporative coolers are not designed to carry away the heat stored in the mass of the materials in the house. They are designed to cool the air and cool the people so the sizing "rule of thumb" is to calculate the volume of the house and divide it by two (the traditional formula for whole house comfort ventilators). So a 2000 square foot house with 8 foot ceilings that has a 16,000 cubic foot volume would require a evaporative cooler with an 8000 cfm capacity.

The technology is evolving as the price of energy and interest in the impact on the planet goes up. There are maintenance issues with the pads, the effects of all the moisture on the system, and the security concerns with leaving windows open. Pushing a lot of air into the attic without adequate relief vents is as much of a concern for evaporative coolers as it is for whole house comfort ventilators. Thinking through all the aspects of the system remains an important design and installation consideration.

Chapter 17
HUMIDIFIERS, DEHUMIDIFIERS, FILTERS, AND VENTILATION ACCESSORIES

Humidity is a major pollutant in homes, and it is carried on air currents. This is not meant to be an exhaustive treatment of dealing with humidification and dehumidification issues in the home. There are many in-depth resources for that. Care should be taken in selecting and installing any system that essentially may be adding a pollutant (moisture) to a home.

SYSTEM HUMIDIFIERS, DEHUMIDIFIERS, AND FILTERS

Humidifiers add moisture to the air. Dehumidifiers remove moisture from the air. That seems sort of obvious, but why the air is too humid or too dry is not always obvious, and adding one of these devices is definitely not always the solution. It is key to the durability of the house and the comfort and health of the occupants that you determine why the air is too dry or too humid before deciding on the best approach to resolving the problem. Not to belabor the point, but moisture is the cause of many indoor air quality problems, and it is not something to be treated casually.

In a cold climate, a house may be too dry because it has too many air leaks. The outside air is very dry relative to the air in the house. If the house is drafty, all that outside air leaking in will definitely make the air uncomfortably dry. The best solution is to tighten up the house, air-seal the leaks. It will save lots of money on the heating bills and is bound to raise the relative humidity in the house. It may get to the point where better mechanical ventilation is required to remove some of the humidity. Adding a humidifier to a central HVAC system can cause serious problems if the house is too leaky because that warm, humid air striking a cold surface will condense, and the moisture is likely to grow mold and that's not good for either the building or the occupants. It is critically important to understand why the house is too dry before adding a humidifier and address the problem rather than the symptom.

And there's the other side of the problem: the house is too humid. The problem might be solved with better ventilation, making sure that there are adequate air changes, that the fans are working properly, that the ducting goes all the way to the outside, etc. Certain areas like basements or rooms housing priceless antiques may need dehumidifiers.

Remember that the house is a system. Make sure you understand the cause of the humidity issue before proceeding to resolve it.

Both heating and cooling systems will dry out the air. A major function of an air conditioning or cooling system is to remove the excess humidity from the air. Adding a humidifier to an air conditioning system is asking the two systems to work against each other: one adding humidity and one taking it away. If the AC system is not effective enough at removing the moisture, there may be system design issues that should be remedied before adding a secondary dehumidifier to the system that would certainly increase the operating cost.

A portable, room humidifier may need to be added to the room of a child that is suffering from winter colds and whose sinuses are laboring with the dry air. Portable devices should be selected carefully and cleaned regularly. Blowing humidity into the air whether it has been aerated acoustically or through an evaporative medium or by some other method (other than boiling) is bound to carry anything else in the water or in the system with it. Ultrasonic systems create a white dust that settles on surfaces in the room. Ultrasonic and impeller humidifiers disperse the greatest amount of microorganisms and minerals into the air. It is important with any humidifier to clean it very regularly and thoroughly.

The ideal humidity levels in the home in winter should vary between 35 percent RH and 50 percent RH. Dust mites in particular enjoy relative humidity above 50 percent. If air conditioning is not needed, a good dehumidifier should be able to keep the humidity under control. Dehumidifiers can be stand-alone units, units attached to the HVAC ducting to dehumidify the air circulating in the house, or units that are used to dehumidify incoming fresh air for ventilation.

The EPA's Energy Star program does rate dehumidifiers. These models are generally 10 to 20 percent more energy efficient than unrated models. Energy Star rated dehumidifiers are qualified by the amount of water they remove per kilowatt-hour of electricity used: 1.20 to 1.80 L/kWh (2.53 pints/kWh to 3.8 pints/kWh) for standard capacity units and greater than 2.5 L/kWh (5.28 pints/kWh) for high capacity units. Energy Star units range in capacity from just over 11 pints per day to more than 144 pints per day.

There are stand-alone units that are available for basement, primary living space, or individual room applications (see Figure 17.1). If the dehumidifier is to be used in the basement, it is important to use a unit that is rated for colder temperatures because the coils can start freezing on many units if the temperature drops below 65°F. Some units have an anti-frost sensor that will cycle the unit off. Many portable units are designed to have collection tanks that need to be manually emptied, although most have gravity drains with an option for a condensate pump.

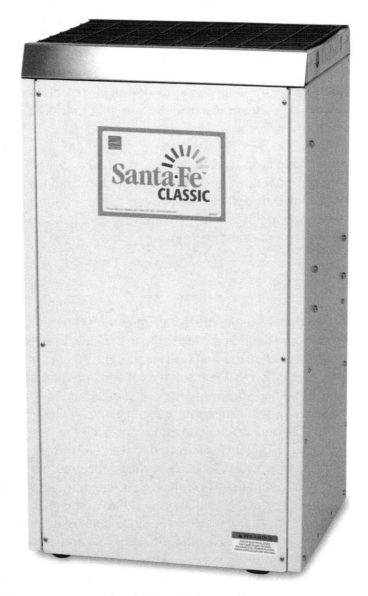

FIGURE 17.1   A system dehumidifier. (Thermastor)

Beyond the decision on how much water you want to remove from the space (which is not always easy to determine), there are energy use and durability (lifecycle) considerations. If you purchase an inexpensive unit that has to be replaced in a year, has it saved you any money? There are calculators online available through Energy Star that will help you determine the energy and dollar savings of Energy Star versus non-Energy Star units.

Whole house dehumidifiers need to be able to effectively circulate their conditioned air throughout the house. Some of them include high efficiency filters and can be installed in-between the outside and the return side of the HVAC system. If the system is taking air from the house, dehumidifying it, and returning it to the house, it can be connected to the supply side of the HVAC system. Units can be installed in the attic or the crawl space. (Wherever they are installed, be sure that there is convenient access to them for maintenance and cleaning.)

Dehumidifiers do add heat to the air. (Air conditioners using the same process add heat as well, but it is expelled to the outside with either a central air conditioning system or a window air conditioner.) The lower moisture level of the air reduces the load on the air conditioning system, making the home more comfortable, and allowing occupants to raise the set point on the thermostat (and save energy).

ROOM AIR CLEANERS

Rather than cleaning all the air moving through an air handler, stand-alone, single-room devices are available in a wide variety of types and sizes. Indoor air quality is a major and reasonable concern, and manufacturers and retailers have jumped on that fear to offer all sorts of approaches to cleaning the air—some valid and some bordering on dangerous.

Negative-ion generators emit negatively charged ions into the air that attach themselves to dust particles, causing the dust particles to have a negative charge. The dust is attracted to positively charged surfaces like walls and ceilings, which serve as massive dust collectors. So they do remove the particles at least temporarily from the air, creating dirty walls and ceilings in the process. They also produce some ozone in the process.

Ozone generators generate ozone that may react with VOCs (volatile organic compounds) in the air and may convert a few to harmless water vapor, carbon dioxide, and oxygen. Ozone generators can also increase the level of VOCs in the house when the ozone reacts with some non-volatile and semi-volatile VOCs. Health Canada concluded in January 1999 that ozone generators pose a risk to the health and safety of the public.[1] The California Air Resources Board states, that "some devices that are advertised as "air purifiers," air cleaners, or ozone generators purposely emit large amounts of ozone, the main component of smog!"[2] They include a continuously updated list of more than 60 manufacturers of devices that have not been tested to the required test protocol regulation. The Environmental Protection Agency describes ozone as, "a toxic gas with vastly different chemical and toxicological properties from oxygen."[3] With all the evidence against them, devices that claim to clean the air, adding ozone in the process, are not an acceptable alternative.

Room filtration systems pull the air in through a filter or series of filters and then push it back out into the room. They come in a wide variety of prices and styles. Keeping the system simple and ignoring all the sales hype is probably the best approach. Look for a system that is made well and well supported.

FURNACE FILTERS

Any ventilation system that brings air in from the outside also brings in the dust, pollen, and other particulates that are floating in the outside air. Filters in HVAC systems do filter the air circulating about the house, but they also protect the mechanical equipment. The equipment doesn't really mind very small particles, particles that are less than 2.5 micron in size. The filters on many HVAC systems remove only larger particles, coarse dust in the 2.5 to 10 micron range. Furnace filters are filters for the furnace not filters for the occupants (unless the system is running all the time).

There are a wide range of filtration systems available including electrostatic, media, pleated, HEPA, and activated carbon air filters (see Figure 17.2). Note that all filtration systems require maintenance. They remove and hold airborne junk! They have no way to get rid of the stuff they collect. As the air passes through the filter and collects, the air passages gradually clog, increas-

1. Health Canada, Consumer Product Safety, "Air Cleaners Designed to Intentionally Generate Ozone (Ozone Generators)," Questions and Answers, http://hc-sc.gc.ca/cps-spc/house-domes/electron/cleaners-air-purificateurs_e.html.
2. California Environmental Protection Agency, Air Resources Board, http://www.arb.ca.gov/research/indoor/o3g-list.htm.
3. EPA, "Ozone Generators That Are Sold as Air Cleaners," http://www.epa.gov/iaq/pubs/ozonegen.html.

FIGURE 17.2   Various filter styles. (Venmar)

ing the efficiency of the filter but blocking the airflow, reducing the efficiency of the air moving system. Until the filter is cleaned or replaced, all the airborne stuff it has collected will remain trapped but still in the air stream.

ASHRAE through Standard 52.2 has developed a filter rating system known as MERV (minimum efficiency reporting value) that simplifies filter ratings (see Table 17.1). The higher the MERV rating the more effective the filter is at removing smaller particles. The period at the end of this sentence is about 400 microns across. Fine human hair is about 40 microns in diame-

TABLE 17.1   MERV filter effectiveness.

MERV Rating	Particle Size		
	0.3–1.0 microns	1.0–3.0 microns	3.0–10.0 microns
1–4	-	-	Less than 20%
5	-	-	20–34.9%
6	-	-	35–49.9%
7	-	-	50–69.9%
8	-	-	70–84.9%
9	-	Less than 50%	More than 85%
10	-	50–64.9%	More than 85%
11	-	65–79.9%	More than 85%
12	-	80–89.9%	More than 90%
13	Less than 75%	More than 90%	More than 90%
14	75–84.9%	More than 90%	More than 90%
15	85–94.9%	More than 90%	More than 90%
16	More than 95%	More than 95%	More than 95%

ter. House dust can range from 100 microns to less than 0.01 microns. Skin flakes fall into the 10 to 25 micron size range, bacteria 0.25 to 10, and viruses in the 0.001 to 0.1 micron range. So if the system uses a MERV 6 filter, it will not be at all effective at removing the small particles, and it will be 35 to 49.9 percent efficient at removing pollen, mold spores, skin flakes, lint, and some animal dander. It's not until you put in a MERV 13 filter that you will be reducing the full range of particulates.[4]

The problem is that most of the higher-level MERV filters increase the resistance to the flow of air through the air handler. It is important to check with the system manufacturer to determine the highest level MERV filter that will work with the system. Unfortunately not all filter manufacturers make it easy to find the MERV ratings on the packages. System manufacturers don't make it easy to determine the maximum filtration level either. That's partially because the performance of an air handler is dependent on its installation. The manufacturer can provide a maximum airflow resistance level, but much of that may be taken up by the ducting and other installation components, leaving an unknown amount of tolerance for the resistance of the filter.

There are a number of test methods other than ASHRAE's that manufacturers may choose to reflect in their advertising. Stating that a filter is 90 per-

4. In most circumstances MERV 11 filters are adequate.

On its Filtrete™ filters 3M has its own rating system called the microparticle performance rating (MPR). In order to address particles between 0.3 and 1.0 micron in size 3M developed the MPR scale. On the MPR scale, 1250 is approximately equivalent to MERV 12, 1000 MPR is approximately equivalent to MERV 11, and 600 MPR is approximately equivalent to MERV 8.

cent efficient without relating what test was used or the size of the particles is essentially useless information. The weight-arrestance test, for example, was developed to measure the efficiency of a filter's ability to capture Arizona road dust over 8 microns in size! This might be helpful to the blower in the HVAC system, but is not very effective at improving health for the occupants of the home.

Filters that are located in the HVAC system will remove particulates from the air as the air circulates through the system and through the house. The filtering efficiency will improve as the air moves around and around the system, passing multiple times through the filter. Filters work best on the inlet side of the air handler with the air being pulled through them. The airflow is more uniform and passes through more of the filter surface. Located on the inlet side, the filter also helps to protect the equipment. If the filter is located on the outlet side, the airflow can be quite turbulent and the particulates not easily removed. When the HVAC system is not running, airborne particles will settle throughout the house where they can't be removed by air filtration. For maximum effectiveness, the air should pass through the filter two to four times per hour. So if it is desirable to use the HVAC system as the whole house filtration system, long, or even continuous, operation of the system is required.

Bringing the ventilation air in on the return side of the air handler allows the outside air to pass through the furnace filter. This system can include its own filtration to further protect the occupants. A "filter-box" can be added to the ductwork that will allow for the use of a reasonably high MERV level filter in the line. These boxes consist of a sheet metal box with duct adaptors on both sides (see Figure 17.3). The box must include the ability to be opened and a filter inserted. The size should be relative to the ducting dimensions as well as a common filter size so that replacements can be easily found. It should be located in such a way that replacing the filter can be done as regularly as the air handler filters.

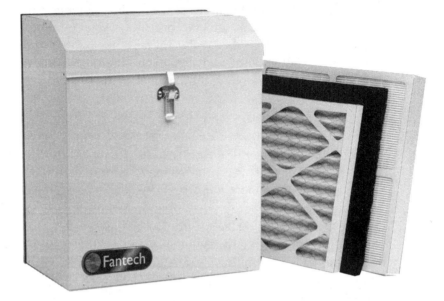

FIGURE 17.3   A HEPA by-pass filter. (Fantech)

Many HRVs and ERVs have only minimal filtration. Filter boxes can be added to the intakes of these devices as well. The additional filtration will not only help the occupants breath better, it will also keep the heat exchanger elements cleaner and more efficient. (It is important that the filters be matched to the capabilities of the equipment. The increased filtration may increase the airflow resistance beyond the capability of the fans.)

HEPA (high efficiency particulate air) filters exceed ASHRAE's 52.2 Standard testing. All HEPA filters must meet a minimum efficiency of 99.97 percent at 0.3 microns. They were originally developed during World War II by the Atomic Energy Commission to filter out deadly plutonium particles in nuclear laboratories. Until recently, they were most commonly used in commercial applications—electronics laboratories and "clean" rooms.

Washable media filters should not be washed multiple times as the moisture in the filter can actually provide an enhanced growth medium for mold and bacteria. Filters that can be replaced and thrown-away are the safest approach.

In residential applications, to keep the frequency of replacement down, most of these systems include pre-filters that remove the largest particles before the air reaches the HEPA filter. They can be installed on the return side of the air handler, being activated each time the blower turns on. The internal blower in the filtration unit draws some of the house air through the filters and then returns it to the HVAC air stream. Without the additional blower in the filter housing, the air would simply bypass the filter, seeking the easiest path through the system. Unfortunately because it only filters part of the air, the overall effectiveness of these "by-pass" filters is very limited.

In most cases, HEPA filtration may be more than is needed in a residential application unless there are especially sensitive occupants in the house, particularly in a tight house where the air is infrequently circulated through the HVAC system. If the HEPA filter is a bypass system that filters 10 to 20 percent of the air moving through the system, if the system were perfectly effective and cleaned only the dirtiest air on each pass, it would take five to 10 complete passes through the system to clean all the air.

Electronic filters need frequent cleaning. Manufacturers recommend four times a year, but in many applications monthly cleaning is required to maintain effectiveness. Although they start out at close to 90 percent efficiency, the plates fill with dust quickly, dropping the efficiency to 40 percent or less in a few days. Because of the high voltage used in these systems, they also produce low levels of ozone that can dissipate quickly if the system is run continuously. If the system includes a humidifier, it should not be located upstream of an electronic filter because the aerosolized moisture will land on the charged plates and reduce the filter's efficiency further.

It's hard not to go for all the marketing hype on filters because occupant health is important, the information is unclear, there will always be confusing claims, and people have different sensitivities.

- The basic furnace filters that you can hold up to the light and see through may be okay for the furnace, but don't do much for the people;
- Use the highest MERV rated filter your furnace air handler can move air through effectively;
- Clean/replace your filter regularly—at least twice a year;
- Seal up the opening where the filter is inserted with a removable medium such as tape or a cover so that the air handler won't suck in non-system air (this will also improve the system's efficiency).

Chapter 18
INDOOR AIR/ENVIRONMENTAL QUALITY CONCERNS

Homes are essentially boxes of air, admittedly fairly complex boxes of air. What they are made of, what we put in them, how we condition them, and how we use them all contribute to the contents of the air inside them (see Figure 18.1). Books, conferences, organizations, and tons of research papers have been created around this subject so this will not be an exhaustive discussion. However, the point of residential ventilation is to improve the quality of the air in the home. If the junk in the air in the house did not affect people's health, then ventilation systems would not be necessary, but, of course, we would be very different creatures. Humans are generally incredibly tolerant of many of the pollutants we put into our air on a regular basis. The greatest impact is on the very young and the old, and infirm and people with compromised immune systems. Some people are very sensitive or allergic to particular pollutants.

Although the impacts of relatively high levels of pollutants are obvious, such as death, the onset of cancer, and increased levels of asthma, there is still not a great deal of information on low level effects, and it is even more difficult to clearly link health effects and ventilation rates. But it is clear that these indoor pollutants do have an impact on health, removing the pollutant from the living space is the most effective approach, and diluting the concentration of the pollutant in the air is the next best step in reducing the potential impact. If there is a pile of excrement on the floor of the living room, it doesn't matter much what kind it is. It is more important to remove it!

There are a number of valuable indoor air quality resources for people who suffer with multiple chemical sensitivity (MCS). The National Center for Healthy Housing (NCHH), the American Lung Association (ALA), and the Healthy House Institute, as well as a wide variety of MCS Websites, are great sources for information.

The most common air contaminants found in buildings:

- Are given off when something is burned. This includes gas stoves, fireplaces and woodstoves, and candles;
- Enter through cracks and holes from the ground (radon and soil gas) or from outside including from an attached garage (see Figure 18.2);
- Are out gassed from building materials and furnishings like cabinets and carpets (formaldehyde);
- Are generated by activities in the house like aerosols and perfumes;
- Are the result of organisms living off the moisture, the materials, and human skin flakes (mold, dust mites, animals, bacteria, viruses);

FIGURE 18.1   Some household pollutants. (Panasonic/Morrissey)

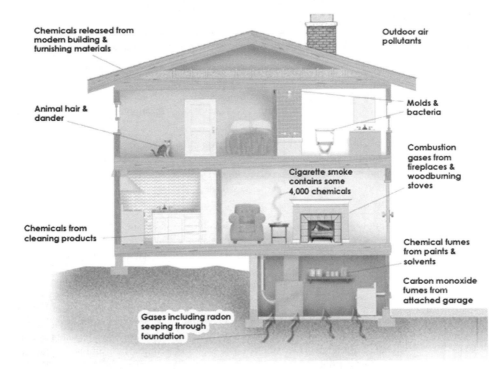

FIGURE 18.2   Pollutant sources. (Panasonic/Morrissey)

- Are the high levels of humidity produced by water generated in the house by human activities, plumbing leaks, moisture and condensation from the outside.

The simplest and most effective ways of reducing and preventing IAQ problems include:

- Don't bring the polluting stuff into the house in the first place. (Verify that the materials used in furniture and carpeting do not include formaldehyde, for example. Don't use unvented kerosene heaters in the house. Using a generator in an attached garage is an invitation to disaster even if the door to the outside is wide open.)
- Isolate any source that is in the house. (Cover or seal asbestos, seal cabinet backs and sides prior to installation, etc.)
- Ventilate near the source of the pollutant. (Bathrooms, kitchens, hobby areas, etc.)

TABLE 18.1   Pollutants.

Pollutant	Typical Source	Equipment & Materials	Ventilation strategy	Occupant maintenance
Radon & Radon daughters	Soil and rock under the house, drawn in through cracks in the foundation	Radon proof construction techniques, EPA's Home Buyer's & Seller's Guide to Radon E-2121	Vent the sump, sub-slab, or basement perimeter crack out through the roof	Maintain fan, periodically verify that it continues to run
Asbestos	Insulation on pipes and old furnaces, also found in siding and "popcorn" ceiling paint	Don't use, cover existing asbestos with a sealant, have removed professionally	Vent the basement or source location	Maintain the ventilation system, don't disturb the asbestos
Formaldehyde and other Aldehydes	Composite wood products like particle board, OSB, foam insulation, paneling, fiberglass insulation	Use low emission materials, seal source materials such as cabinets	Vent the house at a higher rate when introducing these materials; continuous, whole house ventilation	Remove source materials from the house if possible
Suspended combustion particulates	Unvented gas or kerosene appliances; wood stoves & fireplaces; cigarettes, candles; aerosol sprays	Pilotless ignition; proper chimney and combustion air; smoke outside	Ventilating hood; whole house ventilation; provide adequate combustion supply and exhaust	Use range fan; load woodstoves with an updraft; improve insulation & air sealing to limit need for auxiliary heating; read product instructions; smoke outside
Carbon Monoxide (CO)	Gas stoves; unvented heaters; wood stoves and fireplaces (particularly at the end of the burn); attached garages; backdrafting; cigarettes, etc.	Pilotless ignition; proper chimney and combustion air; carefully seal the house/garage interface; sealed combustion appliances; smoke outside	Functional range hood that vents outside; supply correct combustion air; vent the garage (but don't backdraft any appliance in the garage); isolate house intake air from the garage or driveway	Use the ventilation system; maintain the ventilation system; load woodstoves with an updraft; smoke outside, away from intake openings
Carbon Dioxide (CO$_2$)	People; unvented heaters; cigarettes; air from attached garages	Can't remove the people; don't use unvented heaters in inhabited spaces; seal the house/garage interface; smoke outside	Central ventilation system; isolate house intake air from garage or driveway	Maintain the ventilation system and HVAC filters; don't run the car in the garage; keep the door between the house and garage closed
Nitrogen Dioxide (NO$_2$)	Unvented heaters; gas stoves; cigarettes; outside air	Pilotless ignition; don't use in inhabited spaces	Ventilating hoods; smoke outside	Use range fan; improve insulation & air sealing to limit need

Adapted from Brennan & Turner, "Indoor Air Quality: Problems & Solutions," Northeast Sun, June, 1985.

- Use and maintain primary ventilation systems. (Change filters, clean HVAC cabinets (water sitting at the bottom of the drain pans will grow mold or dry up and become dust that gets airborne), remove central humidifiers, etc.)
- Keep the humidity in the house between 30 and 50 percent RH. (This will keep down mold, dust mite, and other micro-organism growth.)

MOISTURE

Moisture is the principal pollutant in the house. It is the source for many other pollutants like mold and dust mites. The good thing is that in comparison to other pollutant sources, it is relatively simple to measure and control. Low cost digital hygrometers can measure the humidity in a room reasonably accurately. Humidity will vary throughout the house and throughout individual rooms. It may be 30 percent RH in the middle of the room and 70 percent RH near a window where it is cooler. Relative humidity or RH is a measure of the amount of moisture the air can hold at a particular temperature. If the air is at 100 percent RH, it can't hold any more moisture. It is at the dew point so that moisture will condense on any surface that is at a lower temperature. A glass of 76°F water in a 75°F room at 100 percent RH will not have condensation on it. Lowering the temperature of the water to 74°F will cause condensation to form. If ice is added to the glass, the temperature of the water will drop and on a very local basis the temperature of the air surrounding the glass will also drop. The temperatures of all the surfaces and all the materials in the house will not be at exactly the same temperatures. Wherever the temperature is below the dew point of the air, condensation will form. The glass of the mirror in the bathroom may be at the same temperature as the rest of the room before you take your shower, but as the RH in the surrounding air increases to 100 percent, the mirror may be cool enough for moisture to form on it. Dew forms on blades of grass because they are completely exposed and cooled by the air moving around them. Any surface in the house or in its structure that is below the dew point at any time will have moisture condense on it. Moisture is sustenance and refreshment for mold and dust mites. When the relative humidity rises above 70 percent, mites can grow in dust without added moisture from our bodies. As the relative humidity rises, dust mites "increase their rate of reproduction, intake of skin scales, and defecation."[1] A level of humidity below 30 percent can cause respiratory problems and dry, cracking skin.

1. May, Jeffrey, My House is Killing Me!, The Johns Hopkins Press, 2001, p. 14.

IAQ stands for indoor air quality. IEQ stands for indoor environmental quality. There are issues that affect the health and comfort of the occupants of the house that are not necessarily floating around in the air. Sound is one of them. As homes get tighter, noise levels go down. Noises that would have been masked by road noise are more obvious and more annoying. IEQ represents the whole system approach to indoor environmental quality.

Some building moisture issues are blatant like a broken water heater tank or an over-filled bathtub. Other moisture issues may be hidden in the walls and develop over an extended period of time. Vinyl wallpaper, for example, is a vapor barrier. In an air-conditioning climate warm, moist air that gets sucked into the building through the wall systems and strikes the back of the wallpaper, can't get through and condenses there. Peeling the wallpaper away will reveal the mold.

Understanding the role of building pressures is critical in keeping moisture out of the building. In a house with a central return, closing bedroom doors at night without adequate pressure relief to bypass the door will cause the bedrooms to increase in pressure and the remainder of the house that is connected to the returns to be depressurized. In air-conditioning climates that depressurization will suck the warm, moist air into the building system. It may not be much, but like a slow drip, it will bring in enough moisture to grow mold and cause serious indoor air quality problems. The opposite effect is also true for a heating climate. Pressurizing the bathroom, for example, may force warm, humid air through any cracks or leaks in the wall system. Mold can also grow in the ducting for the exhaust fan, particularly if it isn't used all that much either because it is too noisy or poorly installed.

Supply-only ventilation systems are not often used in heating climates because of the potential of pressurizing the building and forcing the moisture into the wall system. If the humidity level in the house is low enough (25 percent RH or less), air that does get forced into the wall system won't reach the dew point and condense. But even at 40 or 50 percent RH there can be moisture problems.

For new construction, there are well proven wall construction techniques for every climate that will minimize the potential of hidden mold problems in the building structure.[2] Available materials, building codes, and heating and

2. http://www.eeba.org/bookstore/cat-Builders_Guides-4.aspx.

cooling systems make it mandatory that builders follow these steps. Construction errors can result in massive mold problems, health issues, and a gaggle of lawyers on your doorstep.

Adding a primary, continuously running ventilation system to an existing house requires making sure that the existing HVAC system can handle an additional load. Introducing warm, humid air to an air-conditioned home, will require the AC system to work harder to remove the additional moisture from the air. The addition of the ventilation systems should not be thought of as isolated components but rather as part of the system for the house. Are there ways to reduce the cooling/dehumidification load of the house? These may be structural or mechanical issues. It may require the use of an Energy Recovery Ventilation (ERV) system that removes some of the moisture from the incoming air, transferring it to the outgoing air stream.

A well-installed, spot ventilation system with an inlet located near the pollutant source, such as in a bathroom or kitchen, can remove much of the pollutant before it causes a problem. That's why fume hoods are used in laboratories.

OTHER POLLUTANTS

There are certainly other pollutants that affect indoor air quality. Radon is a radioactive gas that is a natural decay product of radium, which is found in varying quantities in the ground virtually everywhere (see Figure 18.3). Neighboring houses can have very different radon levels. It is an invisible and odorless gas that decays to progeny or daughters that can lodge themselves in the lungs and cause cancer. The Surgeon General has warned that radon is the second leading cause of lung cancer in the United States today, with more than 21,000 annual radon-related lung cancer deaths. Because it is chemically inert, most inhaled radon is rapidly exhaled, but the inhaled progeny lodge in the lungs where they irradiate sensitive cells in the airways, enhancing the risk of cancer.[3] Radon concentrations are generally expressed in picocuries per liter (pCi/L). The average outdoor level is less than 0.5 pCi/L. The average U.S. indoor level is 1.3 pCi/L, but levels over 2000 pCi/L have been mea-

---

3. "EPA Assessment of Risks from Radon in Homes," June 2003, www.epa.gov/radon/pdfs/402-r-03-003.pdf.
    "Model Standards and Techniques for Control of Radon in New Residential Buildings," www.epa.gov/radon/pubs/newconst.html.

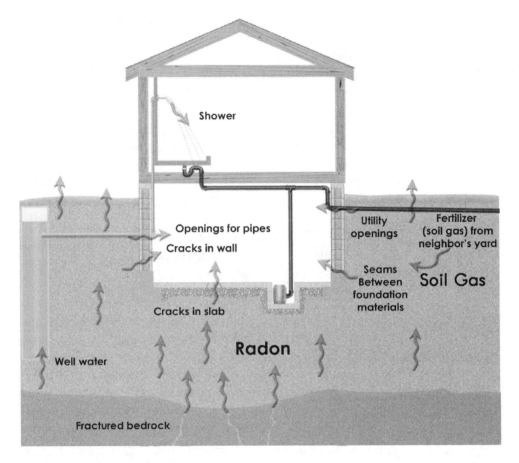

FIGURE 18.3   Radon & soil gas pollution. (Panasonic/Morrissey)

sured. The EPA recommends remediation at levels above 4 pCi/L although the risks of contracting lung cancer from radon at that level is about equal to your lifetime risk of drowning.

Smokers have a much higher risk of getting lung cancer (about 15 times greater than non-smokers) if there are elevated levels of radon in their home. The two factors are additive. If you inhale smoke, you also inhale more particulates of other things in the air. Infants, the elderly, and people with compromised immune systems are at much higher risk with a combination of radon and tobacco smoke in the air.

Small amounts of radon can be released from building materials (including granite counter tops and concrete) and from the water. Generally these levels are low enough and dissipate quickly enough so as not to be a concern. A primary ventilation system running continuously, can remove those pollutants along with the rest of the chemical soup in the air.

As described previously, radon from the ground under the house should be treated as a separate ventilation issue from the rest of the air and pollutant sources in the house. As a part of new construction, many programs require the installation of radon mitigation systems under the slab with a pipe going all the way up through the roof of the house even before the house is tested. (EPA has national and state maps for estimates of radon levels by county. House by house measurement, however, is needed to determine if one particular house has radon-polluted air.[4]) If the system is installed for future use, it should still run all the way up through the roof and it should still be connected to the sub-slab system. It should not be terminated in the attic, and it should not be vented out through a wall. Both of these locations can cause problems with the pollutants flowing back into the house, and even if the piping is clearly marked it can be confused with other piping and employed for other purposes. By venting out through the roof, you can be sure that the pollutants are carried away from the house.

Ground moisture and other soil gases can enter the living space through diffusion, passing through the concrete walls into the living space. Air passing through the soil can pick up moisture and soil gases and move into the living space through small cracks and holes in the foundation. This is particularly true if the basement is under negative pressure. That moisture can lead to mold and the growth of other organisms as well as the increased outgassing of formaldehyde and other VOCs from building materials and furnishings. "Moisture entering the space from the ground is one of the most significant ways water vapor gets into houses. In fact it often dwarfs the amount of moisture generated indoors by the occupants."[5] Keeping the water away from the foundation walls is a structural issue that results in an IEQ issue.

Pesticides and herbicides are poisons. They are formulated to kill creatures and plants. We apply them to our lawns and track them in on our shoes. The toxics are also drawn in through the air moving through the soil. The herbicides spread on the lawn and even your neighbor's lawn can be drawn down through the soil and into your house. Chemicals such as chlordane that used to be injected into the ground around houses to prevent termites have been found in homes years after it was banned. Just like the radon gas in the soil,

4. EPA map of radon zones, www.epa.gov/radon/zonemap.html.
5. Bower, <u>Understanding Ventilation</u>, p. 122.

chemicals spread on the lawn to make it beautiful and chemicals spread around the foundation walls to keep the termites, ants, and rodents away goes into the soil around the house by design. Those gases are as likely as radon to get pulled into the house. Although not as common a health issue as radon, utilizing and maintaining the radon mitigation system for other gases is a great strategy. If you suspect that pesticides and other chemicals are present in your basement, test kits are available for checking for a wide variety of pollutants including nitrogen dioxide, formaldehyde, and a variety of yeasts, bacteria, and fungus.[6]

Carbon monoxide (CO) is an odorless, invisible, deadly gas. It is commonly generated in an incomplete combustion process. As that lovely, romantic fire in the fireplace dwindles down to a few smoldering embers, the chimney begins to cool, and the stack effect, driving the air up the flue becomes weaker and weaker, making it easier for gentle backdrafting to draw the CO into the house.

One of the drawbacks of the new, very efficient combustion appliances is that the temperature of the air going up the chimney is lower, making it easier for these devices to backdraft and draw CO into the house. This makes sealed combustion and powered exhaust systems more necessary.

CO is measured in part per million (ppm). The short-term maximum exposure allowed by the Environmental Protection Agency (EPA) is 9 ppm over eight hours or 35 ppm for one hour. Typical, UL listed CO alarms provide no warning below 30 ppm (see Table 18.2). At 70 ppm residential detectors must alarm after a required time delay. At 100 ppm there will be slight headaches, tiredness, dizziness, and nausea after several hours of exposure. Manufacturers warn that people who may be particularly sensitive to lower levels such as pregnant women, infants, elderly people, and people with compromised immune systems may need devices that provide audible and

TABLE 18.2   UL 2034 CO detector activation levels.

Concentration (ppm)	MUST activate before:	MUST NOT activate before:
30	N/A	8 hours
70	240 minutes	60 minutes
150	50 minutes	10 minutes
400	15 minutes	4 minutes

6. www.healthgoods.com

It is critically important to perform combustion appliance zone (CAZ) testing before and after performing any "air sealing" in the home. Air sealing will reduce the amount of infiltration that would naturally occur, reduce the cost of heating the home, and reduce unpleasant drafts. It may also choke off the combustion air needed for the proper operation of "natural" combustion appliances that rely on infiltration to expel their exhaust fumes.

visual signals for CO levels below 30 ppm. At present, for a CO detector to receive UL-2034 approval the detector must <u>not</u> activate the alarm before the level exceeds 30 ppm for eight hours.

Elevated CO levels in the house are often attributed to malfunctioning furnaces, but it can come from other sources including attached garages. It is often difficult to link the elevated levels to starting the car because the two events may be separated by several hours. The air pressure in the garage should be lower than the air pressure in the house so that any pollutants in the garage flow to the outside rather than into the house.

COMFORTABLE AIR

The goal of any residential ventilation system is to surround the occupants with clean, unpolluted, comfortable air. Outside the house people tolerate all sorts of conditions. We dress for the hot weather and the cold weather, the wind, and the rain. We don't expect the conditions to be constant.

Inside the house, however, we expect to live in a very narrow band of temperature and humidity. We expect our heating, cooling and ventilation systems to keep the air comfortable and safe. There are four factors that must be controlled to surround the occupants of the house with comfortable air: (1) temperature, (2) humidity, (3) air motion and distribution, and (4) the purity of the air or the minimization of the pollutant contents including odors, dust, toxic gases, and bacteria.

ASHRAE has defined "acceptable air quality" as "air toward which a substantial majority of occupants express no dissatisfaction with respect to odor and sensory irritation and in which there are not likely to be contaminants at concentrations that are known to pose a health risk."[7] In the "Handbook of

7. ASHRAE Standard 62.2-2007.

Fundamentals," Chapter 8, ASHRAE describes human comfort as emphasizing "that judgment of comfort is a cognitive process involving many inputs influenced by physical, physiological, psychological, and other processes." It also states that, "Surprisingly, although climates, living conditions, and cultures vary widely throughout the world, the temperature that people choose for comfort under similar conditions of clothing, activity, humidity, and air movement has been found to be very similar."[8] Comfort conditions vary day to day, by age, by the adaption to the surroundings, by perceived outside conditions, by sex, and by seasonal and circadian rhythms. The heating or cooling system must allow occupants to effectively lose enough heat to permit the proper functioning of the metabolic system and yet not lose heat so rapidly that their bodies become chilled. Our systems burn up the food we consume, warming our bodies so that they produce enough heat to keep our temperatures above that of our surroundings. As long as our bodies can effectively dissipate the heat produced at a rate equal to heat production, we remain comfortable.

The body dissipates heat to the ambient air moving past its surface by both conduction and convection, which is why both temperature and air motion are important comfort factors. The skin is also very effective at radiating heat to and absorbing radiant heat from surrounding objects. Our bodies are always giving up moisture to the surrounding air as long as the air around us is not saturated. Heating and evaporation of moisture into the air that enters the lungs also cools the core of the body. For these reasons, controlling the moisture content of the air is obviously an important factor.

The mechanical equipment cannot automatically anticipate and adjust for all these issues and for all the occupants who will have their own momentary criteria, despite the fact that controls are getting increasingly more "intelligent." Occupants like to have control over their environment. If they don't have any control and they don't understand why the system is doing what it is doing, they are likely to believe that it is defective and attempt to manually defeat it.

The heating and cooling equipment control thermal and humidity conditions as well as the majority of air movement through the house. The primary ventilation systems are there to remove pollutants (including humidity) and introduce fresh air into the home. Special purpose ventilation systems are there for cooling, radon and soil gas removal, garage exhaust, attic ventilation, etc. And effective spot ventilation like bathroom and range hood fans

8. ASHRAE Handbook of Fundamentals, 2005, p. 8.1.

remove moisture and pollutants at the source, improve air quality, and govern the fourth factor of indoor comfort.

Although some air motion in a room (in the 15 to 25 feet per minute range) contributes to a sense of the air being fresh, a "draft" or undesired local cooling of the human body is annoying. It causes people to want to raise the temperature or change clothing or location.

To minimize the possibility of drafts caused by the mechanical equipment, the vents (both the supplies and the returns) should be properly located and sized so that the air sweeps through the room effectively but the majority of movement is out of the "occupied areas." This is particularly true where bare skin will be quickly cooled like the face or the hands or the entire torso in a bathroom.

Ventilation air supplied by HRV/ERVs or positive pressure ventilation systems is tempered but should still be delivered out of the occupied area because its temperature is commonly below skin temperature.

It is a particular challenge to supply large volumes of make-up air for clothes dryers and oversized range hoods (as described in Chapter 8). There is a serious energy cost to use electric coils to warm the air up to room temperature. Some make-up air can be supplied behind refrigerators where it will carry away some of the heat from the coils, but this may still cause unwanted drafts across the kitchen floor.

Another factor of air motion is distribution of conditioned air through the house, designing the system so that there is adequate fresh air where occupants spend most of their time, like bedrooms. Pockets of cold air under windows or on outside walls may allow the surface temperature to drop below the dew point, allowing mold to grow. Too much moving air can cause drafts, dry the air, and transport pollutants. Too little moving air can cause pockets where bacteria can grow and spaces that feel stale and uncomfortable.

People spend a very high percentage of their time indoors. It is critically important to their health, safety, and well being that the systems conditioning the air surrounding them be effective and work as a system.

Chapter 19
THE FUTURE OF THE RESIDENTIAL VENTILATION ARTS

Residential ventilation has been around for a long time, and as building science and materials have changed, we have forgotten or chosen to ignore a lot of design features that were developed along the way. Some adventures in airflow are best left in the past (like double envelope houses), but some (like transoms) deserve another look. Clearly understanding the house and the occupants as an integrated and inter-reacting system leads to understanding the potentials of utilizing one system's waste as another system's treasure. Refrigerators, for example, expel waste heat. Could that waste heat be used to preheat incoming ventilation air? What would that mean to the operation of the refrigerator? How would it work when the incoming air was warm? Such questions inevitably lead to more questions. Unlike in the past, we now have tools to model innovative changes before we invest in them, before we build them, before we use our friends and neighbors as guinea pigs! We also have the tools to measure their performance once they have been built and installed.

There is general agreement that mechanical ventilation is an important factor in homes (see Figure 19.1). There is a great deal of discussion on what the ventilation rate should be. How much air should be moved? Should it be constant or intermittent? Should the same levels apply to all homes or should there be regional rates? Certainly other parts of the world have different ventilation requirements, most of which are higher than the U.S. rates.[1] As concerns for indoor pollutants increase, it is likely that ventilation rates will increase. Until there are better approaches to distributing the ventilation air throughout the house, higher rates may force more air to move throughout the building. As distribution and control systems improve, rates may come down again.

The ability to measure performance will make the greatest changes in ventilation systems in the future. The problem in writing about the future is that

1. McWilliams, Sherman, "Review of Literature Related to Residential Ventilation Requirements," INIVE EEIG, 2007.

each word ahead of this one was the future at some point. And now it's the past. Computer modeling is helping us to determine the effects of air distribution throughout homes based on assumed removal of contaminants. That in turn may lead to new approaches to fresh air delivery and pollutant removal systems.

FIGURE 19.1   A distributed exhaust system, "SmartSense." (Broan)

Measurement tools can help us determine how well installed systems are working. Knowing what a ventilation system does in the test laboratory is very nice and a great place to start, but how is it actually working now that it has been installed? It is important to know how well the system is working when it has been installed and how well it continues to work. Building codes and green building programs may describe what airflows need to be, but if they don't actually measure them, how can they know that is truly what is happening in the house? Codes may be changed to require installed system measurements by building inspectors. That will mean that inspectors and installers will need access to affordable, accurate, and simple to use measuring tools. Measuring low flows of installed systems is difficult due to building and site pressures. Hopefully we will be able to figure out how to take those variations into account and be able to make accurate and repeatable measurements. Perhaps manufacturers could build flow indicators into their equipment so that changes in performance could be monitored over time. That would mean that homeowners would have to have some understanding of their ventilation system and how to maintain it. That sort of "feedback" has proven to be an exceptionally good modifier of behavior.

Homeowner understanding and knowledge must change if there is going to be an impact on how we all use energy. Understanding smoke detectors and CO detectors, changing furnace filters, setting thermostats, and turning off lights in unused rooms, are all occupant behavior issues that can have a major impact on how much energy we all use. When the heating system was a massive fireplace in the middle of the house, everyone knew when it needed more wood. Central heating systems were relegated to the basement. Grilles and radiators were refined into artistic elements or hidden under boxes. Ventilation systems were made so quiet that it is impossible to know that they are running without some sort of indicator light. Home occupants just assume that the air around them will be magically warm or cool and clean and comfortable. Many don't know where their thermostat is, what it is for, or how to set it. Should we continue to make these things less obvious and more automatic or should we force people to get more involved with the operation of their homes? Perhaps equipment manufacturers should do more to make their products more user friendly, and less dependent on contractors to change filters, adjust settings, or replace batteries. Maybe we need nagging voices in machines that say, "You haven't changed my filter in 12 months. It's not good for you to breathe this air. Wipe your feet when you come in the house, and change my filter. Do it now before you forget!" (On the other hand, maybe voices in machines is not such a great idea. Remember the nagging seat belt voices in cars?)

If we can get the energy load down low enough, maybe we can figure out a way to effectively integrate all the comfort conditioning systems together, combining heating, cooling, and air conditioning in an energy efficient package (see Figure 19.2). If the building needs less energy for conditioning, it may need less energy for conditioning air delivery and stale air removal. Smaller systems may allow for the use of smaller ducts and lower velocity airflows and more energy efficient delivery systems.

Some companies are already refining the earth conditioning idea that was pursued with "earth tubes." The ground-air heat exchangers line the tubes with a proprietary silver particle-enabled antimicrobial inner layer that inhibits the growth of bacteria.[2]

Disaggregating the ventilation air may also lead to improved systems. Range hoods should be designed to provide their own make-up air. If people want to install those huge commercial hoods, they need to install huge, commercially sized, make-up air systems as well. Furnaces and water heaters should all provide their own combustion air. That would eliminate all back-drafting issues and protect people from carbon monoxide poisoning. Central vacuums, fireplaces, wood and pellet stoves could supply their own make-up air systems as well. Clothes dryers would work better if they drew in their own supply air. (It would be a challenge to create an effective heat exchanger that could return the waste heat from the dryer back into the living space while avoiding the lint build-up issues.) By compelling product manufacturers to think through their complete system process, fewer burdens would be placed on the system that conditions and cleans the air for the primary ventilation system.

Do we really need to have a ventilation system that continuously delivers and removes the same amount of air all the time? A continuously operating system is simple and has less that can go wrong with it. It doesn't depend on operator knowledge to make adjustments. But from an energy efficiency standpoint, the ventilation system only needs to be working at a time and at a rate to get its job done.

A radon/soil gas mitigation system, for example, runs 24 hours a day, seven days a week sucking the soil gas out from under the house. But are the levels of soil gas and radon constant? Could the system be controlled so that the extraction rate varied in relation to the conditions? There are a lot of com-

2. http://na.rehau.com/0EEEFF8538D268D18525756E004A1FD7_3A809C0C98EF7B D1C125710A00572EB5.shtml.

plexities to accomplish that, but it is something for the future to think about. What about garage ventilation? Does that flow rate need to be constant or could it be varied in relation to pollutant and pressure levels? Could the system "learn" when homeowners generally start and remove their cars, turn the system on in anticipation, anticipate when they are coming home, and make sure that system runs long enough to get rid of all the gases? Should cars have CO sensors on them that would activate the ventilation system in the garage if the air surrounding them became polluted or should they just shut themselves off so that the people in the house wouldn't die?

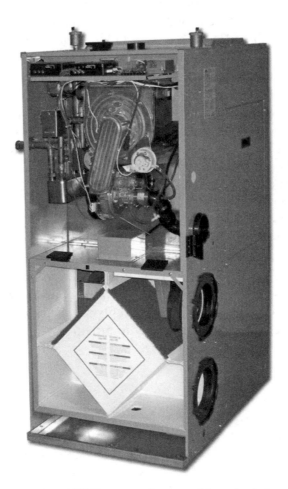

FIGURE 19.2   HVAC and HRV systems integrated in a single housing "matrix." (New York Thermal)

Ventilation controls can be designed that recognize occupant levels in the house, counting people coming in and going out or monitoring the level of CO_2 and increasing the ventilation rate when a party is going on. (These have been available for awhile but generally priced out of the range of residential applications.) Integrated systems are available that cause bathroom fans to "talk" to each other, cycling from one to the next to make sure there is some circulation in the house. Motion detectors and movement sensors delay off and cycling timers are all presently available for bathroom fans. Mixed gas sensors are available that measure a combination of gases to boost or activate HRVs and ERVs.

There is certainly room for improvement in "termination fittings" like exterior hoods. Unconditioned air is not effectively stopped at the exterior of the building. Hoods are incredibly restrictive to airflow—they are made with too sharp a turn in order to minimize how visible they are on the outside of the house. Perhaps they should be made into an architectural element instead of trying to hide them. We used to stick television antennas on the top of our homes. A well-designed exhaust vent would not be better than that. What about combining systems so that there wouldn't be so many penetrations and hoods? In multi-family buildings there are often three penetrations in the walls per apartment—bathroom exhaust, kitchen exhaust, and unit make-up air. If there are 100 units in the project, that's 300 hoods on the outside of the building.

There will undoubtedly be many, less obvious changes in ventilation design—new fan blades that are quieter and more efficient perhaps fans without blades at all, better, more efficient motors, even fans that have neither blades nor motors! There will be better, more efficient closure systems and better controls—a myriad of tiny tweaks that all add up to more efficient systems. Even in an area as seemingly mundane as residential ventilation, things keep changing and hopefully improving.

Appendix A
ABBREVIATIONS

ACCA	Air Conditioning Contractors of America
ACH	Air changes per hour
ACH_{50}	Air changes per hour at 50 Pascals of depressurization
ACH_N	Natural air changes per hour
bhp	Brake horsepower
BRE	Building Research Establishment
ASEF	Apparent sensible effectiveness
ASHRAE	American Society of Heating, Refrigerating, and Air-Conditioning Engineers
BPI	Building Performance Institute
BTL	Building tightness limit
Btu	British thermal unit
CAZ	Combustion appliance zone
ccf	One hundred cubic feet
CDD	Cooling degree days
CEC	California Energy Commission
cfd	Computational fluid dynamics
c.f.	Cubic foot
cfm	Cubic feet per minute
CFM_{50}	Cubic feet per minute at 50 Pascals of depressurization
cps	Centipoise
CO	Carbon monoxide
CO_2	Carbon dioxide
CSA	Canadian Standards Association
dB	Decibel
dBA	Decibels measured on the "A" scale
DCV	Demand-controlled ventilation
DD	Heating degree days

ΔP	Pressure differential
DTL	Depressurization tightness limit
ECM	Electronically commutated motor
ELA	Effective leakage area (U.S.)
EPA	Environmental Protection Agency
EqLA	Equivalent leakage area (Canadian)
EER	Energy efficiency ratio (related to SEER)
ERV	Energy-recovery ventilator
ESL	Energy systems laboratory
ETL	Electrical testing labs (Intertek)
EQ	Environmental quality
fpm	Feet per minute
fms	Flow measuring station
GAMA	Gas Appliance Manufacturers' Association
HEPA	High efficiency particulate filter
H_2CO	Formaldehyde
H_2O	Water
HRAI	Heating, Refrigeration, and Air-Conditioning Institute of Canada
HRV	Heat recovery ventilator
HVAC	Heating, ventilating, and air-conditioning
HVI	Home Ventilating Institute
IAQ	Indoor air quality
IEQ	Indoor environmental quality
IBC	International Building Code
IECC	International Energy Conservation Code
in. w.g. (or i.wc.)	Inches of water on a pressure gauge or inches of water column
IMC	International Mechanical Code
IPMC	International Property Maintenance Code
IRC	International Residential Code for one- and two-family dwellings
kWh	Kilowatt-hour
L/s	Liters per second
L/h	Liters per hour
LEED	Leadership in Energy and Environmental Design

m^3/s	Cubic meters per second
m.c.	Moisture content
MCS	Multiple chemical sensitivity
MERV	Minimum efficiency rating value
MET Lab	Maryland electrical testing
MPR	Microparticle performance rating (3M Filtrete filters)
MVC	Minimum ventilation capacity
NBC	National Building Code of Canada
nfa	Net free area
NO_x	Nitrogen oxide
NL	Normalized leakage
OTL	Overall tightness limit
Pa	Pascal
PAV	Powered attic ventilator
pCi/L	PicoCurie per liter
ppm	Parts per million
psi	Pounds per square inch
psia	Pounds per square inch absolute
psig	Pounds per square inch gauge
PSV	Passive stack ventilation
Q	Airflow
RESNET	Residential Energy Efficiency Network
RH	Relative humidity
rpm	Revolutions per minute
RSS	Reference sound source
SEER	Seasonal energy efficiency ratio
SLA	Specific leakage area (ELA/floor area)
SMACNA	Sheet Metal and Air Conditioning Contractors National Association
sp	Static pressure or system pressure
sq. ft.	Square foot
sq. in.	Square inch
SRE	Sensible recovery efficiency
TD or ΔT	Temperature differential
TP	Total pressure
TRE	Total recovery efficiency

UL	Underwriter Laboratories
USGBC	United States Green Building Council
VOC	Volatile organic compound
VP	Velocity pressure
W	Watt
w.g.	Water gauge
Wh	Watt-hour
WRT	With respect to

Appendix B
USEFUL VALUES AND MULTIPLIERS

Multiply:	By:	To obtain:
Btu	0.29305	Watt-hours
Cubic feet per minute (CFM)	0.471947	Liters/second (L/s)
CFM	0.028317	Cubic meters/min (m³/min)
CFM	1.699	m³/h
Foot	0.3048	Meters
Foot	304.8	mm
Feet per minute (fpm)	0.00508	m/s
Gas		
Cubic foot (natural gas)	1,028	Btus
Therm	100,000	Btus
Gallon of propane	91,600	Btus
Horsepower	42.42	Btus/minute
Horsepower	0.7457	Kilowatts
Horsepower	746	Watts
Inch of water gauge (in.w.g.) (68°F)	248.64	Pascal
Inch of mercury (in. hg)	3,386	Pascal
Kilowatt-hours	3412	Btus/hour
Oil		
Gallon	139,000	Btus (100% efficiency)
Meters/second (m/s)	200	fpm (more accurately 196.9)
Watts	3.413	Btus/hour
Watts	0.00134	Horsepower

Appendix C
GLOSSARY

The following terms are defined relative to the content of this book. Some of them have other meanings. For example "defrost mechanisms" are used in refrigerators as well as in HRVs.

A

Absolute humidity: The weight of water in a given volume of air, commonly expressed in pounds of water per pound of dry air or grains of moisture per cubic foot. See also relative humidity.

Accidental pressure: A difference in air pressure between the indoors and the outdoors induced by mechanical devices where the primary purpose is not ventilation.

Accidental ventilation: The unintentional movement of air into and out of a house caused by accidental pressures.

Air barrier: Material used to block the flow of air into a building.

Air change rate or air changes per hour (ACH): The number of times in one hour that the volume of air in a house is replaced with outdoor air, sometimes expressed at 50 Pa pressure difference (ACH_{50}) between the indoors and outdoors.

Airflow grid: A device that mounts inside a duct (often temporarily) that can be connected to an air-pressure gauge or manometer to sample a cross section of the airflow through a duct to determine the volume of flow.

Air handler: A cabinet for a conditioning system containing a blower that moves air through a system of heating/cooling ducts.

Air leakage: The uncontrolled exchange of air between the interior and exterior environments through unintentional openings in the thermal envelope of the building. See also infiltration and exfiltration.

Airtight construction: A building technique that reduces most openings for unintentional airflow and results in a house with a small ELA. (Equivalent Leakage Area.) A key component of a Zero Energy House (ZEH).

Air-to-air heat exchanger: A ventilation device designed to be balanced, capable of transferring heat (and sometimes moisture) between two air streams without merging the stream. See also heat recovery ventilator.

Ambient temperature: The temperature that surrounds an object on all sides.

Anechoic chamber: A room designed to attenuate sound.

Apparent sensible effectiveness (ASEF): A measurement useful in determining the temperatures of air streams in a heat recovery ventilator. The ASEF includes the effect of motor heat gain, cross-leakage gain, and casing gain. It is usually numerically higher than sensible recovery efficiency of the HRV.

Aspect ratio: The ratio of the width to the depth, particularly of a duct.

Axial fan: A type of air moving device that has propeller-like blades, often used for window fans and moving air between spaces with little resistance.

B

Backdraft damper: A device that allows air to flow in only one direction.

Backdrafting: Complete reversal of flow in the chimney of a fuel fired appliance, usually due to negative pressures indoors. Combustion products are forced back down the chimney into the living space.

Balanced ventilation: A ventilation strategy that results in the house experiencing a neutral pressure with an equal mechanical flow of air moving into and out of the space.

Balancing: The process of adjusting the flows of a ventilation system so that the house experiences a neutral pressure.

Balancing damper: A device that is mounted in the ducting that allows the ventilation system to be balanced. These generally include a "fixing" method that locks the damper in place once balance has been achieved.

Barometric damper (or draft regulator): A draft control device intended to stabilize the natural draft in an appliance by admitting room air into the venting system.

Biological pollutants: Air particulates and pollutants such as mold and pollen, that either are or once were alive or the byproducts of metabolism.

Blower door: A large fan that mounts temporarily in a doorway used to pressurize or depressurize a house to evaluate its tightness.

Breathing zone: See head space.

British thermal unit (Btu): The heat energy required to raise the temperature of 1 pound of water 1°F, used in calculating building conditioning loads.

Building envelope: The exterior portion of a structure that keeps the weather out.

Building tightness limit (or building airflow standard [BAS], minimum ventilation level [MVL], minimum ventilation rate [MVR]): Terms describing a house tightening limit used for ensuring adequate air quality for the occupants of the house.

Burner: A device for the introduction of fuel into the combustion zone for ignition.

Natural draft burner: A burner not equipped with a fan or blower.

Fan assisted burner: A burner with combustion air supplied by a mechanical device such as a blower at sufficient pressure to overcome the resistance of the burner only.

Forced draft burner: A burner with combustion air supplied by a mechanical device such as a blower at sufficient pressure to overcome the resistance of both the burner and the appliance.

Bypass: A pathway or hole through a building, usually hidden within the structure, that bypasses the air boundary of the building envelope.

C

Capital cost: The installed or first cost of equipment (i.e., the equipment itself, incidentals, and labor.) See also operating cost.

Carbon dioxide (CO_2): A colorless, odorless gas released in exhaled breath, or by a combustion process.

Carbon dioxide sensor: A device that senses the carbon dioxide concentration in the air (usually in ppm, parts per million) and can be used to control a ventilation system based on human occupancy of a space.

Carbon monoxide (CO): A colorless, odorless, poisonous gas released during the incomplete combustion of carbon-containing fuels that in low concentrations causes headaches and nausea and flu-like symptoms and in high concentrations can cause coma and death.

Carbon monoxide detector: A device that senses carbon monoxide in the air and sounds an alarm when the concentration reaches a certain point.

Carcinogen: A substance that may cause cancer.

Centrifugal fan: A ventilating fan that allows the spinning blades to throw off the air by centrifugal force.

Central exhaust ventilation: See exhaust-only ventilation.

Central supply ventilation: See supply-only ventilation.

Chase: A vertical space created to house pipes or ducts. A chase can be open all the way from the basement to the attic.

Chimney: A vertical structure or conduit that encases at least one flue and usually carries combustion byproducts out of a house. See also ventilating chimney.

Chlordane: A persistent chlorinated hydrocarbon pesticide used for the treatment of termites that is noticeable because of its odor.

Circulation fan: The blower in the air handler that draws the return air in from the house and pushes the conditioned air back to the rooms.

Coanda effect: The tendency of an air stream or fluid jet to stay attached to an adjacent curved surface when directed at a tangent to that surface.

Coaxial vent: A venting system consisting of an inner pipe to exhaust air within an outer pipe, drawing in replacement air. Used for both combustion systems and heat exchangers.

Combustion air: Air that enters a house specifically to be used for satisfactory combustion of a fuel. See also ventilation air.

Combustion byproducts: Gases and particulates released during the burning of a fuel. Includes inert gases that are part of the air, but not excess air.

Comfort zone: The range of effective air quality conditions including temperature and humidity over which a majority (>50 percent) of adults feel comfortable.

Condensation: The change of a gas or vapor into a liquid, accompanied by the release of heat; the opposite of evaporation.

Conditioned space: The part of a house, within the building envelope, that is designed to be maintained at a comfortable temperature and humidity.

Conductance (thermal): The rate of heat flow through a unit area of a material per unit of temperature difference. Designated in a "U" value. The lower the "U" value the slower heat moves through the material, the better the insulation value.

Conduction (thermal): The process of heat transfer moving through a material from the warmer side to the cooler side.

Continuous duty: An electrical safety rating for motors, not the same as continuous use. Most motors are rated for continuous duty.

Continuous use: A rating that a fan is capable of running for an indefinite period of time, not the same as continuous duty. Not all motors are rated for continuous use.

Convection (thermal): The transfer of heat that takes place within moving gases and liquids. Moving warm air is continuously replaced by a flow of cooler air.

Cubic feet per minute (cfm): Free area in square feet times the face velocity in feet per minute (fpm).

D

Damper: A movable device, motorized or manually adjustable or reliant on gravity, used to vary or control the airflow in a duct. See also backdraft damper and modulating damper.

Decibel (dB): A logarithmic unit of measurement used to express sound intensity.

Defrost mechanism: A device used in HRVs and ERVs to melt ice that builds up in the core in cold climates.

Degree day: A unit of measurement used to estimate fuel consumption and heating or cooling costs based on temperature and time. For heating this is generally set at 65°F (18°C). If the temperature drops below 65°F (18°C) by 1° for an hour, that would be 1 degree hour. Multiplying by 24 would be one degree day.

Dehumidistat: A control device that can be used to activate a ventilation or dehumidification system that makes contact as the relative humidity rises. Used to decrease the humidity in the space. See also humidistat.

Delta pressure (ΔP): The difference in pressure from one space to another. ΔP will also be expressed in relation to or with respect to (WRT) another space.

Delta temperature (ΔT): The difference or change in temperature.

Demand-controlled ventilation (DCV): The process of automatically supplying air to, and removing air from, a space whenever needed by the occupants, often associated with a CO_2 control.

Depressurization: When the air pressure inside a house is less than the atmospheric pressure outside the house. The rate at which air leaves the house exceeds the rate at which air enters the house.

Desiccant: A drying agent that absorbs moisture, used in some energy-recovery ventilators to enhance moisture transfer between air streams.

Design temperature: The temperature used for sizing a heating or cooling system.

Dew point: The temperature at which air is saturated with moisture (100 percent relative humidity) and below which condensation will occur.

Diffuser: An outlet grille designed to spread out the airflow in a room for proper mixing.

Diffusion: The migration of molecules of a gas or a vapor (or a liquid) from an area of high concentration to an area of low concentration.

Dilution: The mixing of fresh air into polluted air to reduce the concentration of pollutants; also applies to liquids.

Dilution air: Air that is drawn together and mixes with combustion byproducts prior to their being expelled from a house through a natural draft chimney.

Direct vent (sealed combustion): A type of fuel-fired appliance that takes its combustion air directly from the outdoors and ejects its combustion gases directly back to the outdoors through an independent sealed vent. The exhaust gas temperatures of these systems are usually low enough to allow the use of plastic vent piping.

Displacement ventilation: A method of moving air through a room, pushing the fresh air in at the floor level and allowing it to rise through the space, generally only used in specialized applications.

Distribution: Movement of air throughout the rooms of a house through ducting or through the rooms themselves.

Downdraft: An exhaust system that draws the air down and out of the building, often applied to kitchen exhaust fans.

Draft: The pressure difference that causes a current of air or gases to flow through a flue, chimney, or space. Also an uncomfortable breeze or current of air in a room. See also natural draft.

Draft hood or draft cap: A draft control device having no moveable parts that may be built into a combustion appliance or vent connector that is designed to reduce the effect of stack action by drawing dilution air into the flue.

Dry bulb temperature: The temperature of the air indicated by an accurate thermometer. The common measurement of temperature, but referred to "dry bulb" to distinguish it from wet bulb temperature.

Drying potential: The ability of a substance to dry after it becomes wet.

Duct: A conduit for conducting air to intended spaces, made from a variety of materials and in a variety of shapes.

Duct, branch: A conduit for carrying air to or from only one grille or register.

Duct, trunk: A conduit for carrying air to or from multiple grilles or registers.

Duct board: A fiberglass board material with an aluminum-foil facing on one side, used to construct ducts. Affectionately called "fuzzy-duct."

E

Earth tube: A pipe through the ground for moving air from the outdoors into the living space, used for tempering incoming air.

Effective leakage area (ELA): The size of a special, nozzle shaped opening that a house would have if all the small random holes between the indoors and the outdoors were combined into one hole. The number is used in the U.S. and is based on a pressure of 4 Pa. It is the same concept as equivalent leakage area, but is calculated differently.

Effective length: The length of a duct or system path equal to the total of the actual duct length and the equivalent lengths of all the fittings. Used in system design.

Electrically (or electronically) commutated motor (ECM): An ultra high efficiency, programmable brushless DC motor using a permanent magnet rotor and a built-in inverter. ECM motors are electrically efficient, especially at low speeds. They usually include their own speed control.

Electrostatic air filter: An air filter made of plastic materials that capture particulate pollutants using static electricity.

Electrostatic precipitator: An air filter that uses a high voltage to cause particulates to become electrically charged and cling to metal plates having an opposite charge.

Energy recovery ventilator (ERV): An air-to-air exchanger that recovers both latent heat and sensible heat.

Enthalpy: The total amount of heat contained in air that is the sum of the sensible heat and the latent heat.

Entrain: To draw in—used to describe the ability of an airflow to carry particulates.

Equivalent length: The length of straight duct that would have the same resistance to airflow as a fitting. It is expressed in terms of feet of ducting of the same diameter of the fitting, which would have the same resistance to airflow.

Equivalent leakage area (ELA): The area of a single, square-edged hole that a house would have if all the individual random holes between the inside and the outside were combined into one hole. This number is primarily used in Canada and is based on a pressure of 10 Pa. It is the same concept as effective leakage area, but is calculated differently.

Evaporation: The changing of a liquid to a gas or vapor, by adding latent heat. See also condensation.

Exfiltration: Uncontrolled air leaving a house through random holes in the structure moved by building pressure differentials. See also infiltration.

Exhaust air: The air mechanically leaving a house through a ventilation system and not reused. See also supply air.

Exhaust grille: A grille through which "used" air leaves a room. See also supply grille.

Exhaust-only ventilation: A mechanical system that depressurizes a building in order to change the air, relying on gaps in the building shell to make up for the air removed.

Exchange rate: The speed at which indoor air is replaced with outdoor air.

F

Fan: A device that moves air.

Filtration: The process of extracting pollutants from the air.

Flame rollout: Flames pushed outside a combustion chamber as a result of backdrafting.

Flat plate core: A device that is used in some HRVs or ERVs to transfer heat (and sometimes moisture) between two air streams without mixing the two streams.

Flue: An enclosed conduit, pipe, or passageway for conveying combustion gases out of a building; sometimes known as a stack.

Flue gases: A combination of combustion gases and excess air.

Forced draft: A mechanical system that uses a fan to inject air into the combustion chamber of a combustion appliance.

Formaldehyde: A gas found in indoor air, often used as a binder for building materials, consisting of one carbon atom, two hydrogen atoms, and one oxygen atom.

Free area: See net free area (NFA).

Fresh air: Air that has not been polluted.

Fresh-air duct: A conduit used to guide unpolluted air into a building.

Furnace: A flue connected, space-heating appliance that uses warm air as the heating medium, commonly including an air handler and the provision to attach return and supply ducting.

G

General ventilation: Distributed fresh air supply and return for the health and comfort of the building's occupants.

Grille: A duct covering on a wall or ceiling through which air moves to or from the conditioned space. Grilles are generally fixed and non-adjustable.

Grill: A grating that supports comestibles cooking over a fire.

H

Head space: The breathing zone in a room. In a residence it usually extends from a few inches to about 6 feet off the floor.

Heat-recovery ventilator (HRV): A ventilation device capable of transferring heat between two airstreams without blending the streams. See also air-to-air heat exchanger and energy recovery ventilator (ERV).

HEPA filter: A very high-efficiency particulate air filter, often used in hospitals and laboratories. A HEPA filter is supposed to remove 99.97 percent of 0.3 micron particulates from the air flowing through it and 100 percent of the particulates larger than 1 micron.

Humidistat: A control device that activates a ventilation or humidification system that makes contact as the relative humidity falls. Used to increase the humidity in the space. See also dehumidistat.

Humidity: See absolute humidity and relative humidity.

Hygrometer: A device that measures relative humidity.

I

Inches of mercury (in. hg. or hg.): A unit of atmospheric pressure measurement or a height of a column of mercury.

Inches of water gauge (in. wg or wg): A unit of atmospheric pressure measurement or a height of a column of water.

Induced draft: A system utilizing a fan or blower often located in the flue to exhaust air from the combustion chamber of a combustion appliance.

Infiltration: Uncontrolled air entering a house through random holes in the structure moved by building pressure differentials. See also exfiltration.

In-line fan: A ventilating fan made to attach a duct on both the inlet and the outlet sides.

Intentional ventilation: Purposeful air movement into and out of a building in a regular way, usually caused by a fan moving air through intentional openings.

Ionizer: A device that generates ions (usually negatively charged) ostensibly for cleaning the air to mimic natural ionization caused by lightning.

J

Jump duct: A short duct connection between two rooms that "jumps" over the separating wall through which air can move when pressure imbalances occur. Also called a jump-over, jumper, or transfer duct.

K

Kilowatt hour (kWh): A unit for electrical energy measurement equal to 1,000 watts of power consumed over an hour.

L

Latent heat: The energy required in the change of state without a change in temperature or the amount of heat that must be removed from air to change the water vapor from a gas to a liquid or a liquid to a gas. See also enthalpy and sensible heat.

Local exhaust ventilation: A localized ventilation system near the source of air pollutants. See also general ventilation and spot ventilation.

M

Magnahelic gauge: An analog pressure-measuring device.

Make-up air: Outdoor air that enters a house to replace air that is exhausted from the house. It does not include air entering the house as combustion air or to replace air lost through exfiltration.

Manometer: A gauge for measuring air-pressure differences.

Media filter: A pleated filter that relies on a fibrous material, usually polyester or fiberglass, to physically strain particulates out of the air. See also electrostatic air filter and electrostatic precipitator.

MERV rating: A filter rating system developed by ASHRAE via Standard 52.2 (minimum efficiency reporting value). The higher the MERV rating, the more effective the filter is at removing smaller particles.

Micron: A millionth of a meter.

Mildew: Fungus or mold that grows primarily on surfaces.

Minimum ventilation capacity (MVC): In the CSA F326 Standard this is a capacity set by room count (5 L/s per room except the master bedroom at 10 L/s, and 10 L/s for a basement area exceeding two-thirds of the total basement area. The same as total ventilation capacity.

Mixed-gas sensor: A device that senses oxidizable gases in the air.

Modulating damper: A motorized closure system whose opening can be adjusted to regulate a volume of airflow.

Mold: A slimy or powdery growth caused by fungi. Molds require oxygen, moisture and a food source such as cellulose to grow.

Motion sensor: A device that senses movement that can be used to control a ventilation system by assuming room occupancy.

Multi-port ventilator: A ventilation device having a number of ports from running ducts to or from different rooms. See also single-port ventilator.

N

Natural draft: The negative pressure in a chimney caused by rising warm combustion byproducts often referred to simply as draft. This flow of gases relies on thermal buoyancy to vent the combustion products. The upward force must be greater than any resisting forces (including negative pressure) in the building envelope.

Natural pressure: The difference in air pressure between the indoors and the outdoors induced by natural phenomena such as wind, stack effect, and diffusion.

Natural ventilation: Random air movement into and out of a house caused by natural pressures through intentional openings like windows and unintentional cracks and holes.

Negative ions: Negatively charged atoms or groups of atoms that will cling to oppositely charge surfaces.

Negative pressure: See depressurization.

Net free area (nfa): The amount of unobstructed open area of a grille, inlet, or outlet. HVI certified products actually measure the flow through the opening rather than estimating the area blocked by the louvers. Sometimes known as the "K" factor.

Neutral pressure: Air pressure that is neither positive or negative with respect to another space.

Neutral pressure plane: The point of a house that experiences neutral pressure as a result of stack effect.

O

Olf: A unit of odor measurement equal to the amount of body odor produced by one person performing normal activities in one day.

Operating cost: The cost of operating equipment including electrical costs, conditioned air costs, and maintenance costs. See also capital cost.

Outgassing (off-gassing): The release of volatile gases from the surface of a solid material as a part of aging, decomposition, or curing.

Ozone (O_3): An unstable form of oxygen (O_2) consisting of three oxygen atoms that can be produced in small amounts by electric motors, electrostatic precipitators, ionizers, etc.

Ozone generator: A device that creates ozone designed to react with air pollutants to neutralize them.

P

Partial-bypass filter: A filter that removes some contaminants from the air by extracting a portion of the return airflow, filtering it, and then returning it to the primary flow.

Particulate: A microscopic fragment of a solid or droplet of a liquid that is suspended in the air.

Parts per million (ppm): A unit of measurement commonly used to express the concentration of pollutants in the air.

Pascal (Pa): A metric unit of pressure measurement useful in diagnosing houses.

Passive ventilation: A non-mechanical general-ventilation strategy that utilizes natural or accidental pressures that involves air movement through intentional inlets and outlets.

Permanent-split-capacitor motor: A type of electric motor that includes a primary winding and an auxiliary winding wired in series with a capacitor.

Pesticides: Chemical compounds formulated to exterminate living creatures.

PicoCurie per liter (pC/l): A measurement unit of the concentration of radon.

Pitot tube: A device for measuring total pressure, static pressure, and velocity pressure within a duct.

Plenum: An element of the air-moving conduit where the air moving to or from the air-handling device is gathered.

Pounds per square inch (psi): A unit of pressure measurement. One psi equals 144 pounds per square foot.

Pressure drop: The static pressure loss caused by airflow through a duct, filter, or heat or energy exchanger core.

Pressurization: When the air pressure inside a house is greater than the atmospheric pressure outside the house. This occurs when the rate of air entering the house exceeds the rate at which air is leaving.

Primary ventilation system: The principal air moving system (usually mechanical) that circulates air through a building to control the overall air quality to ensure the health of the occupants and the building.

Psychometric chart: A chart that is used by engineers to graphically determine the moisture content of air at different temperatures.

R

Radon: A naturally occurring carcinogen often released as a radioactive gas from soil and rocks.

Radon daughters: The natural radioactive decay particulates of radon, some of which release harmful radiation and can lodge in the lungs.

Recirculate: To move air through a space without adding fresh air or removing polluted air.

Recirculating range hood: A range hood that has no duct, recirculates the air moving through it, and is not connected to the outdoors.

Register: A duct termination fitting that can be adjusted to regulate the amount or direction of airflow.

Relative humidity (RH): The amount of moisture in the air, usually expressed as a percentage, compared to the maximum amount of moisture that air at that temperature can contain. See also absolute humidity.

Return air: Air moving from the living space that is drawn back to an air handler to be reconditioned.

Ridge vent: An opening at the ridge of the house usually covered by a cap that allows the attic air to flow out. The air flows in through vents located in the soffits. See also soffit vents.

Rotary core: A slowly rotating wheel that is used in some air-to-air heat exchangers to transfer heat and moisture between two air streams.

S

Sealed combustion: A system used in some furnaces, boilers, fireplaces, and water heaters that is immune to pressure imbalances in a house because it draws combustion air into a combustion chamber from the outdoors and expels combustion byproducts to the outdoors within totally enclosed pipes and chambers.

Sensible heat: The amount of heat involved in raising or lowering the temperature of air, not including the heat required to cause water vapor to change state (e.g., from a gas to a liquid): heat that can be "sensed." See also enthalpy and latent heat.

Sensible recovery efficiency (SRE): A measurement of heat recovery that does not include latent that is useful in comparing the amount of energy passed between airstreams in a heat-recovery ventilator. It is corrected for the effect of motor heat gain, defrost energy, cross leakage gain, and other effects like casing gain. It will usually be lower than apparent sensible effectiveness of the HRV.

Shaded pole motor: A type of motor whose starting torque is provided by a copper ring called a "shading coil" that delays the phase of magnetic flux.

Soffit vents: Intentional openings in the soffits that allow air to flow in and upward to the ridge vent and out of the attic, ostensibly to keep the roof surface cooler.

Sone: A linear unit of sound measurement used to express low level sound intensities. Two sones are twice as loud as 1 sone.

Sound attenuator: A sound muffler.

Source control: The principle of controlling the polluting materials in the living space.

Specific heat: The quantity of heat required to change the temperature of a substance by one degree.

Spillage: A situation where some combustion byproducts spill into the living space rather than go up the chimney due to insufficient draft.

Spot ventilation: Ventilation for the direct removal of air pollutants near a source. See also local-exhaust ventilation and primary ventilation system.

Stack effect: The natural phenomena of warm air exerting pressure on cooler air that results in warm air rising.

Static pressure: The amount of outward pressure exerted against the walls of a duct or airway, created by the friction and impact of air as it moves.

Supply air: The air entering a house through a ventilation system or HVAC system. See also exhaust air.

Supply grille: A grille through which air is delivered to a room. See also exhaust grille.

Supply-only ventilation: A positive pressure mechanical ventilation system that pressurizes the building and relies on natural leakage to establish flow.

T

Temperature difference (TD or ΔT): The difference in temperature between two volumes of air.

Tempering: The conditioning of outdoor air so it will more closely match the temperature and humidity of the indoor air of the building it is entering.

Thermal boundary: The boundary that thermally separates the interior, conditioned space of a building from the exterior environment.

Through-the-wall vent: A intentional opening in an exterior wall through which a controlled quantity of air is allowed to move. See also trickle ventilator.

Total recovery efficiency (TRE): The energy recovered in an air-to-air heat exchanger that compensates for the fan energy and cross-leakage.

Total pressure: A non-existent pressure that expresses the sum of static and velocity pressures.

Transfer grille: A term used to describe the opening between rooms that allows for the equalization of pressure. See jump duct.

Trickle ventilator: An intentional opening between the interior and the exterior of a building intended to equalize the pressure and provide for the transfer of polluted and fresh air.

Turbulent flow: The irregular swirling of air particles in an air stream that offers more resistance than laminar flow.

V

Vapor barrier: A material used to block the flow of vapor through a structure.

Ventilating chimney: A vertical structure, pipe or conduit designed to passively carry ventilation air into or out of a house.

Ventilation: The process of intentionally supplying air to, removing air from, or moving air around a house, most often with a fan, for the health and comfort of the occupants. See also controlled ventilation.

Ventilation air: Air that intentionally enters a house for the purpose of ventilating. See also combustion air.

Volatile organic compound (VOC): Molecular compounds containing carbon that easily evaporate, often released from building materials and found as contaminants in indoor air. See also outgassing.

W

Wet bulb temperature: The temperature measured by the wet-bulb thermometer of a psychrometer.

Wetting potential: The ability of a substance to sequester moisture after it has dried out. See also drying potential.

Whole house fan or whole house comfort ventilator: A fan used to exhaust air from a space specifically for the purpose of enhancing the natural cooling effect of moving air through a house.

Appendix D
REFERENCES

Anderson, Bruce (with Michael Riordan), <u>The Solar Home Book</u>, Cheshire Books, 1976.

ASHRAE Guideline 24-2008, "Ventilation and Indoor Air Quality in Low-Rise Residential Buildings," American Society of Heating, Refrigerating, and Air-Conditioning Engineers, Inc., www.ashrae.org, 2008.

ASHRAE Handbook 2008, <u>HVAC Systems and Equipment</u>.

ASHRAE Handbook 2005, <u>Fundamentals</u>.

ASHRAE Standard 62.2-2007, "Ventilation and Acceptable Indoor Air Quality in Low-Rise Residential Buildings," American Society of Heating, Refrigerating, and Air-Conditioning Engineers, Inc., www.ashrae.org, 2007.

Axley, James W. , "Residential Passive Ventilation Systems: Evaluation and Design," AIVC Technical Note 54, 2001, p. 158, Code TN 54.

Begley, Jr., Ed, <u>Living Like Ed</u>, Clarkson Potter Publishers, 2008.

Bleier, Frank, <u>Fan Handbook, Selection, Application and Design</u>, McGraw-Hill, 1998.

Bower, John, <u>Understanding Ventilation</u>, The Healthy House Institute, 1995.

Brennan, Terry, and Turner, William, "Indoor Air Quality Problems & Solutions," *Northeast Sun*, June 1985.

Brennan, Terry, and Clarkin, Mike, "A Guide to Ventilating a House," New York State Energy Research and Development Authority.

Brumbaugh, James E., <u>HVAC Fundamentals</u>, Wiley Publishing, 2004.

California Department of Health Services, "Health Hazards of Ozone-Generating Air Cleaning Devices," www.cal-iaq.org/o3_fact.htm, 1998.

Chen, Dr. Bing; Wang, Dr. Tseng-Chan; Maloney, Dr. John; Ennenga, Prof. James; and Newman, Mark, "Measured Cooling Performance of Earth Contact Cooling Tubes," Progress in Solar Energy, American Solar Energy Society, 1984.

CMHC, "Garbage Bag Airflow Test," About Your House, available on the CMHC Website.

CMHC, "Garage Performance Testing," Innes Hood at The Sheltair Group, January 2004.

Cummings, James B. and Withers, Jr., Charles R., "Unbalanced Return Air in Residences: Causes, Consequences, and Solutions," Florida Solar Energy Center, *ASHRAE Transactions*, 2006.

Dumont, R.S., and Makohon, J.T., "Characterization of Volatile Organic Emissions from Building Materials for Indoor Environment Assessment," *EEBA Journal*, Winter/Spring 1997.

Edwards, Rodger, Handbook of Domestic Ventilation, Elsevier Butterworth-Heinemann, 2005.

Emmerich, Steven J.; Gorfain, Joshua E.; Huang, Mike; and Howard-Reed, Cynthia, "Air and Pollutant Transport from Attached Garages to Residential Living Spaces," NIST, NISTIR 7072, 2003.

Emmerich, Steven J. and Persily, Andrew K., Multizone Modeling of Three Residential Indoor Air Quality Control Options, NISTIR 5801, March 1996.

Fugler, Don, "Is it Worth Putting in a Better Furnace Filter?," *Home Energy Magazine*, May/June 2000.

Fugler, Don, "The Impact of Vacuuming," *IAQ Applications*, ASHRAE, September 2004.

Greiner, Thomas H., and Schwab, Charles V., "Carbon Monoxide Exposure from a Vehicle in a Garage," Thermal Envelopes VII, 1998.

Greiner, Thomas, "The Silent Killer: Carbon Monoxide," Iowa State University, January 1998.

Greiner, Thomas, "Wholistic Approaches to CO," Affordable Comfort Conference, Indianapolis, 2005.

Grimsrud, David T., "Intermittent Ventilation and Standard 62.2," *IAQ Applications*, ASHRAE, Fall 2004.

Hayes, Vicky, and Shapiro-Baruch, Ian, "Evaluating Ventilation in Multifamily Buildings," *Home Energy*, July/August 1994.

Havrella, Raymond A., Heating, Ventilating, and Air Conditioning Fundamentals. Gregg Division of McGraw-Hill Book Company, 1981.

Health Canada, "Air Cleaners Designed to Intentionally Generate Ozone (Ozone Generators)—Questions and Answers," 2007.

Hickory Consortium, "Analysis of Current Practice and Advanced Ventilation Strategies for Residential Buildings," Building America Initiative, Deliverable 2.A.1, February 1999.

HRAI, Bean, Robert, Residential Indoor Air Quality Awareness, Heating, Refrigeration, and Air Conditioning Institute of Canada, 2009, www.hrai.ca.

HRAI, Residential Mechanical Ventilation, Heating, Refrigeration, and Air Conditioning Institute of Canada, 2004, www.hrai.ca.

Indiana Community Action Association, "Weatherization Assistance Program Indiana Field Guide, Chapter 5, Airflow," 2007 www.incap.org.

Janssen, John E., "The History of Ventilation and Temperature Control," *ASHRAE Journal*, September 1999.

Jennings, Burgess, Environmental Engineering—Analysis and Practice. International Textbook Company, 1970.

Kadulsky, Rick, "Attached Garages and Indoor Air Quality," *Solplan Review*, March 2001.

Karg, Rick, "Survey of Tightness Limits for Residential Buildings," July 2001, for the Chicago Regional Diagnostics Working Group, www.karg.com/btlsurvey.htm.

Kelley, Mark, "Ventilation System Performance in Energy Crafted Homes," *Building Science Engineering*, November 1995.

Krigger, John, and Dorsi, Chris, Residential Energy, Saturn Resource Management, 2004, www.srmi.biz.

Lawrence, F. Vinton, "Natural Energy Systems in Traditional Tibetan Buildings," *Sunstone Design*, 1986.

Lstiburek, Joseph, "Understanding Attic Ventilation," *Building Science Digest 102*, 2006.

Manclark, Bruce, "Oversized Kitchen Fans—An Exhausting Problem," *Home Energy Magazine*, January/February 1999.

May, Jeffrey C., My House is Killing Me!, The Johns Hopkins Press, 2001.

McKone, T.E., and Sherman, Max H., "Residential Ventilation Standards Scoping Study," LBNL Technical Assistance for PIER, LBNL-53800, October 2003.

McWilliams, Jennifer, "Review of Airflow Measurement Techniques," Energy Performance Buildings Group, LBNL-49747, 2002.

McWilliams, Jennifer, and Sherman, Max, "Review of Literature Related to Residential Ventilation Requirements," INIVE EEIG, 2007.

Offerman, Francis, et al, "Window Usage, Ventilation, and Formaldehyde Concentrations in New California Homes," Indoor Air 2008 Conference, August 2008.

Oikos, "Range Hoods Affect Indoor Air Quality," www.oikos.com/esb/53/range-hood.html.

Palmiter and Brown, The Northwest Infiltration Survey, ECOTOPE, Aug. 5, 1989, Prepared for the Washington State Energy Office.

Parker, Danny S., and Sherwin, John R., "Performance Assessment of Photovoltaic Attic Ventilator Fans," Florida Solar Energy Center (FSEC), FSEC-GP-171-00.

Pekkinen, Jorma S., "Thermal Comfort and Ventilation Effectiveness in Commercial Kitchens," *ASHRAE Journal*, July 1993.

Raymer, Paul H., "Make Room for the Caddy," *Home Energy Magazine*, March/April 2005.

Rizzi, Ennio A., <u>Design and Estimating for Heating, Ventilating, and Air Conditioning</u>, Van Nostrand Reinhold Company, 1980.

Rizzuto, J., <u>An Investigation of Infiltration and Indoor Air Quality</u>, NYSERDA 90-11, 1989.

Rizzuto, J., Nitchke, Traynor, Wadach, Clarkin, and Clark, <u>Indoor Air Quality, Infiltration and Ventilation in Residential Buildings</u>, NYSERDA 85-10, 1985.

Rose, William B., and TenWolde, Anton, "Venting of Attics & Cathedral Ceilings," *ASHRAE Journal*, October 2002.

Rudd, Armin, <u>Ventilation Guide</u>, Building Science Press, 2006.

Seppänen, Olli, Fisk, William J., and Mendell, Mark J., "Ventilation Rates and Health," *ASHRAE Journal*, August 2002.

Sherman, Max, and Walker, Iain, "Can Duct-Tape Take The Heat?," Energy Performance of Buildings Group, Lawrence Berkeley National Laboratory, University of California, LBNL-41434.

Sherman, Max H., and Walker, Iain, "Energy Impact of Residential Ventilation Standards in California," Environmental Technologies Division, LBNL-61282, 2007.

Sherman, Max H., "On the Valuation of Infiltration towards Meeting Residential Ventilation Needs," Environmental Energy Technologies Division, Lawrence Berkeley National Laboratory, University of California, LBNL-1031E.

Sonne, J., and Parker, D., "Measured Ceiling Fan Performance and Usage Patterns: Implications for Efficiency and Comfort Improvement," ACEEE Summer Study on Energy Efficiency in Buildings, Vol. 1, 1998.

Steven Winter Associates, "Evaluation of Three Ventilation Systems in Chicago Homes, Final Report for Field Evaluation of PATH Technologies," March 2006.

Tooley, John, "Duct Diagnostics," Affordable Comfort Conference, March 1995.

Vent-Axia, System Calculator, www.vent-axia/knowledge/handbook/calculator.asp.

Walker, Iain, "Garbage Bags and Laundry Baskets—Homemade Air Flow Diagnostic Tools Get Professionally Tested," *Home Energy Magazine*, November/December 2003.

World Health Organization (WHO), "Development of WHO Guidelines for Indoor Air Quality: Dampness and Mould," Working group meeting, October 2007.

Yberg, Ingvar, <u>Indoor Air—The Silent Killer</u>, Svensk Ventilation, 2004.

Yberg, Ingvar, <u>Shortness of Breath—A Handbook About the Air in Our Homes</u>, Svensk Ventilation, 2005.

Appendix E
ORGANIZATIONS AND RESOURCES

This section includes trade associations, places to learn, and organizations that are working with energy efficient tight housing that requires mechanical ventilation. Please note that because the world changes so quickly, addresses, phone numbers, Websites, and even names may have changed.

PLACES TO LEARN MORE

Affordable Comfort, Inc.
ACI presents some of the most informative building science conferences around the country.
www.affordablecomfort.org
32 Church Street
Suite 204
Waynesburg, PA 15370
Phone: 800-344-4866

Air Infiltration and Ventilation Center (AIVC)
The AIVC offers industry and research organizations technical support aimed at optimizing ventilation technology. AIVC offers a range of services and facilities, including comprehensive database on literature standards, guides, technical notes, and ventilation data.
www.aivc.org
INIVE eeig
Lozenberg 7
B-1932 Sint-Stevens-Woluwe
Belgium
Phone: +32 2 655 77 11

Alliance for Healthy Homes
A national, nonprofit, public interest organization working to prevent and eliminate hazards in homes that can harm the health of children, families, and other residents.
www.afhh.org
50 F St. NW, Suite 300
Washington, DC 20001
Phone: 202-347-7610

BP Consulting
Online building science training courses.
www.bpconsulting.org
141 Oak St.
Ballston Spa, NY 12020
Phone: 518-309-3415

BuildingGreen, LLC
www.buildinggreen.com
An organization committed to improving the environmental performance and reducing the adverse impact of buildings. (Publisher of Environmental Building News and GreenSpec)
122 Birge Street Suite 30
Brattleboro, VT 05301
Phone: 802-257-7300

Building Performance Institute (BPI)
BPI offers nationally-recognized training, certification, accreditation, and quality assurance programs.
www.bpi.org
107 Hermes Road, Suite 110
Malta, NY 12020
Phone: 877-274-1274

Canadian Mortgage and Housing Corporation (CMHC)
CMHC is Canada's national housing agency whose mission is to enhance the quality, affordability, and choice of housing in Canada. CMHC has an excellent resource library that includes a great deal of data and articles on residential ventilation and IAQ issues for both Canada and the U.S.
www.cmhc-schl.gc.ca
700 Montreal Road
Ottawa, Ontario, Canada K1A 0P7
Phone: 613-748-2000

Energy and Environmental Building Association (EEBA)

EEBA provides education and resources to transform residential design, development, and construction industries to deliver energy efficient and environmentally responsible buildings and communities.
www.eeba.org
6520 Edenvale Boulevard, Suite 112
Eden Prairie, MN 55346
Phone: 952-881-1098

The Engineering Toolbox

A wide array of engineering information, tools, charts, and conversion factors.
www.engineeringtoolbox.com

Enterprise Green Communities

The first national green building program developed for affordable housing.
www.greencommunitiesonline.org
10227 Wincopin Circle, Suite 500
Columbia, MD 21044
Phone: 410-715-7433

Delmar Cengage Learning

Online learning for the building trades and construction and numerous other fields.
www.informationdestination.cengage.com

National Center for Healthy Housing

NCHH is a nonprofit based in Columbia, MD dedicated to creating safe and healthy homes for children.
www.nchh.org
10320 Little Patuxent Parkway, Suite 500
Columbia, MD 21044
Phone: 410-992-0712

National Comfort Institute

www.nationalcomfortinstitute.com
P.O. Box 2090
Sheffield Lake, OH 44054
Phone: 800-633-7058

Northeast Sustainable Energy Association (NESEA)

www.nesea.org
50 Miles Street
Greenfield, MA 01301
413-774-6051

North American Technician Excellence (NATE)
www.natex.org
2111 Wilson Blvd., Suite 510
Arlington, VA 22201
Phone 877-420-NATE

Oikos
Iris communications produces and supports the Oikos Websites that are dedicated to sustainable and energy-efficient construction, green building news, and green building products.
www.oikos.com
P.O. Box 6498
Bend, OR 97708-6498
Phone: 541-317-1626

Residential Energy Services Network (RESNET)
RESNET supports the HERS (home energy rating system) index of home efficiency that is recognized by financial institutions for federal tax incentives and the EPA's Energy Star™ Program.
www.natresnet.org
P.O. Box 4561
Oceanside, CA 92052-4561
Phone: 760-806-3448

United States Department of Energy
Energy efficiency and renewable energy state energy office contacts. This is a great place to get more information about your local programs.
http://apps1.eere.energy.gov/state_energy_program/seo_contacts.cfm

United States Environmental Protection Agency (EPA)
Instigator, manager, and purveyor of the Energy Star programs including Energy Star for Homes Indoor Air Plus. Includes a listing of all the ventilating fans that are classified as Energy Star at:
http://www.energystar.gov/index.cfm?c=manuf_res.pt_vent_fans.
1200 Pennsylvania Ave. NW
Washington, DC 20460
Phone: 888-782-7937
www.energystar.gov

U.S. Green Building Council (USGBC)
Instigator, manager, and purveyor of the LEED programs.
www.usgbc.org
Washington, DC
Phone: 800-795-1747

Yestermorrow Design/Build School

Yestermorrow inspires people to create a better, more sustainable world by providing hands-on education that integrates design and craft as a creative, interactive process.

www.yestermorrow.org
189 VT Route 100
Warren, VT 05674
Phone: 802-496-5545

TRADE ASSOCIATIONS

Air-Conditioning, Heating and Refrigeration Institute (AHRI)

A national trade association of manufacturers of central air conditioning, warm-air heating, and commercial and industrial refrigeration equipment.

www.ahri.org
2111 Wilson Blvd, Suite 500
Arlington, Virginia 22201
Phone: 703-524-8800

Air Conditioning Contractors of America (ACCA)

A national trade association of heating, air conditioning, and refrigeration systems contractors. Publishers of Manual J and Manual D.

www.acca.org
2800 Shirlington Road Suite 300
Arlington, VA 22206
Phone: 703-575-4477

Air Diffusion Council (ADC)

An organization that was formed to promote the interests of the manufacturers of flexible air ducts and related air distribution equipment. Detailed flexible ducting installation materials.

www.flexibleduct.org
1000 E. Woodfield Road Suite 102
Schaumburg, IL 60173
Phone: 847-706-6750

American Gas Association (AGA)

The AGA develops residential gas operating and performance standards and maintains system performance and statistical records.

www.aga.org
151400 North Capitol Street, N.W.
Washington, DC 20001
Phone: 202-824-7000

American Society of Mechanical Engineers (ASME)

A nonprofit technical and educational association offering educational and training services and conducting technology seminars and onsite training programs.

www.asme.org
Three Park Ave
New York, NY 10016
Phone: 800-843-2763

Heating, Refrigeration, and Air Conditioning Institute of Canada (HRAI)

Skill-Tech Academy—Ventilation training and IAQ resources (not just for Canada!)

www.hrai.ca
2800 Skymark Avenue, Building 1, Suite 201
Mississauga, Ontario, Canada L4W 5A6
Contact HRAI for further information at 1-800-267-2231 or 905-602-4700 ext. 246

Home Ventilating Institute (HVI)

HVI is a nonprofit trade association representing residential ventilation products sold in North America. HVI certifies the performance of residential ventilation products through a very carefully controlled testing process.

www.hvi.org
1000 N Rand Rd.
Suite 214
Wauconda, IL 60084
Phone: 847-526-2010

Indoor Air Quality Association, Inc.

A nonprofit, multi-disciplined organization dedicated to promoting the exchange of indoor environmental information through education and research for the safety and well-being of the general public.

www.iaqa.org
12339 Carroll Ave.
Rockville, MD 20852
Phone: 301-231-8388

Heating, Air Conditioning & Refrigeration Distributors International (HARDI)

HARDI is a national trade association of wholesalers and distributors of air conditioning and refrigeration equipment.

www.hardinet.org

3455 Mill Run Drive Suite 820

Columbus, OH 43026

Phone: 614-345-4328

Toll Free: 888-253-2128

National Association of Home Builders (NAHB)

NAHB is a trade association that helps promote the policies that make housing a national priority.

www.nahb.org

1201 15th Street, NW

Washington, DC 20005

Phone: 202-266-8200

National Association of the Remodeling Industry (NARI)

NARI's core purpose is to advance and promote the remodeling industry's professionalism, product, and vital public purpose.

www.nari.org

780 Lee Street Suite 200

Des Plaines, Illinois 60016

Phone: 800-611-NARI (6274)

Sheet Metal and Air Conditioning Contractors National Association (SMACNA)

An international trade association of union contractors who install ventilating, warm-air heating, air-conditioning, and air handling equipment.

www.smacna.org

4201 Lafayette Center Dr.

Chantilly, VA 20151

Phone: 703-803-2980

TEST, CERTIFICATION, PROFESSIONAL, AND STANDARDS ORGANIZATIONS

Air Movement and Control Association International, Inc. (AMCA)

A trade association of the manufacturers, wholesalers, and retailers of air movement and control equipment. AMCA publishes test standards and certifies the performance of ventilation products.

www.amca.org
30 West University Drive
Arlington Heights, IL 60004
Phone: 847-394-0150

American Society of Heating, Refrigerating, and Air-Conditioning Engineers (ASHRAE)

ASHRAE is an international professional organization concerned with the advancement of the science and technology of heating, ventilation, air conditioning, and refrigeration. ASHRAE maintains Standard 62.2 for residential ventilation.

www.ashrae.org
1791 Tullie Circle NE
Atlanta, GA 30329
Phone: 800-527-8400

Bodycote Testing Group

A product testing group that is used by the Home Ventilating Institute members to test HRVs and ERVs. They also test and verify the performance of many other products.

www.bodycotetesting.com
2395 Speakman Drive
Mississauga, ON   Canada L5K 1B3
Phone: 613-634-9307

Canadian Standards Association (CSA)

CSA is the Canadian equivalent of UL. They test products to both Canadian and U.S. standards, issue the CSA mark, and require product maintenance inspections.

www.csa-international.org
178 Rexdale Boulevard
Toronto, ON Canada M9W 1R3
Phone: 866-797-4272

International Code Council (ICC)
A membership association dedicated to building safety and fire protection and the developer of codes used to construct residential and commercial buildings, including homes and schools.
www.iccsafe.org
500 New Jersey Avenue, NW
Washington, DC 20001-2070
Phone: 888-422-7233

Texas A&M Energy Systems Laboratory (ESL)
The test and research facility for the heating, ventilation, and air conditioning industry. Used by HVI for the certification of ventilation products.
http://esl.eslwin.tamu.edu/riverside-lab.html
Texas Engineering Experiment Station
Riverside Campus—Building 6502
3100 S.H. 47 Bryan, TX 77807
Phone: 979-845-6334

Underwriter Laboratories, Inc. (UL)
UL is an independent, nonprofit product safety testing and certification organization.
www.ul.com
333 Pfingsten Road
Northbrook, IL 60062
Phone: 8467-272-8800

Appendix F
EQUIPMENT SOURCES

ILLUSTRATOR

Morrissey Designs
Source of great illustrations—quick, accurate, and fun to work with.
www.paddydesigns.com
Paddy Morrissey
681 Peralta Ave.
Berkeley, CA 94707
Phone: 510-527-8009

FAN AND VENTILATION EQUIPMENT

Note that many of these companies are members of the Home Ventilating
Institute and have certified products. Contact information for all the HVI
related companies can be found at:
For member companies with certified products:
http://www.hvi.org/manudist/memberswithHVICP.html
For member companies without certified products:
http://www.hvi.org/manudist/mmbrswoutHVICP.html
For associate members:
http://www.hvi.org/manudist/assocmembers.html
For non-members with certified products:
http://www.hvi.org/manudist/nonmmbrswithHVICP.html

Active Ventilation Products, Inc.
Manufacturer of a variety of attic and roof ventilation products.
www.roofvents.com
P.O. Box 1521
Newburgh, NY 12551-1521
Phone: 800-247-3463

Airia Brands, Inc.
Manufacturers of the Lifebreath HRVs and ERVs.
www.lifebreath.com
511 McCormick Boulevard
London, ON N5W 4C8
Phone: 519-457-1904

Air-King
Manufacturer of a wide range of residential ventilation products.
www.airkinglimited.com
820 Lincoln Ave.
West Chester, Pa 19380
Phone: 610-436-1454

American Aldes
Manufacturer of a wide range of unique ventilation products.
www.americanaldes.com
4521 19th Street Court East Suite 104
Bradenton, FL 34203
Phone: 800-255-7749

Broan-NuTone
Manufacturer of a variety of consumer products including numerous ventilation fans.
www.broan-nutone.com
926 West State Street
Hartford, WI 53027-1098
Phone 262-673-4340

Exhausto
The manufacturer of the premier chimney top fan systems.
www.us.exhausto.com
1200 Northmeadow Pkwy. Suite 180
Roswell, GA 30076
Phone: 770-587-3238

Fantech
Manufacturer of in-line fans and air-to-air heat exchangers.
www.fantech.net
1712 Northgate Blvd
Sarasota, FL 34234-2112
Phone: 800-747-1762

Heyoka Solutions, LLC

Products for ventilation and building analysis, product development, education, and the home of the author of this book. Heyoka Solutions develops ventilation products such as the Comfort Stream whole house comfort ventilator for other companies.
www.HeyokaSolutions.com
19 Howes Lane
Falmouth, MA 02540
Phone: 866-389-8578

Imperial Air Technologies, Inc.

Manufacturer of the Greentek and Imperial lines of ventilation products.
www.imperialgroup.ca
480 Ferdinand Boulevard
Dieppe, NB E1A 6B9
Phone: 888-724-5211

Lipidex (AirCycler)

Manufacturer of ventilation control systems.
www.aircycler.com
411 Plain Street
Marshfield, MA 02050
Phone: 781-834-1600

New York Thermal, Inc.

Makers of the integrated HVAC and HRV system "The Matrix."
www.ntimatrix.com
30 Stonegate Drive
Saint John, NB, Canada E2H 0A4
Phone: 800-688-2575

Nu-Air Ventilation Systems, Inc.

Manufacturer of a variety of air-to-air heat exchangers and controls.
www.nu-airventilation.com
16 Nelson Street
Windsor, NS B0N 2T0
Phone: 902-798-2261

Panasonic Ventilation Systems

A source of extremely quiet, efficient, and well made fans and ventilation related products.
http://www.panasonic.com/business/building-products/ventilation-systems/
One Panasonic Way 1H-3
Secaucus, NJ 07094
Phone: 866-292-7292

Quality Aluminum Products
Manufacturer of aluminum siding, soffits, fascia, trim coil, and rain carrying products.
www.qualityaluminum.com
Flat Rock
Michigan, U.S.
Phone: 734-783-0990

Renewaire, LLC
Manufacturer of air-to-air heat and energy exchangers and bath fans.
www.renewaire.com
4510 Helgesen Drive
Madison, WI 53718
Phone: 800-627-4499

Soler & Palau U.S.A.
Manufacturer of in-line fan products.
www.solerpalau-usa.com
6393 Powers Ave.
Jacksonville, FL 32256
Phone: 800-961-7370

Universal Metal Industries
Manufacturer of a wide range of residential ventilation products including bath fans and range hoods.
www.umiphx.com
800 W. Grant
Phoenix, AZ 85007-2409
Phone: 800-875-3654

Venmar Ventilation
Venmar makes heat exchangers, range hoods, and other residential ventilation products.
www.venmar.ca
550 Lemire Blvd
Drummondville, (Quebec) Canada J2C 7W9
Phone: 800-567-3855

Ventamatic Ltd
Manufacturer of the NuVent residential ventilation products and numerous other fans.
www.bvc.com
100 Washington Street
Mineral Wells, TX 76068
Phone: 800-433-2626

Vent Axia
A British ventilation manufacturer that includes exceptional system design information on its Website.
www.vent-axia.com
Fleming Way
Crawley, West Sussex RH10 9YX
United Kingdom
U.S. Phone: 800-735-7026

FRESH AIR MAKE-UP SYSTEMS

Electro Industries, Inc.
Tempered fresh air make-up systems.
www.electromn.com/gen/makeup_air.htm
2150 West River Street
Monticello, MN 55362
Phone: 800-922-4138

EP Sales
Make-up air systems, particularly EPS systems.
http://www.epsalesinc.com/products.php
7878 12th Ave. Sout
Bloomington, MN 55425
Phone: 952-854-4400

Shelter Supply
Make-up air systems as well as other ventilation products.
www.sheltershupply.com
151 East Cliff Road, Suite 30
Burnsville, MN 55337
Phone: 800-762-8399

Titon Inc.
British manufacturer of "trickle vents."
www.titon.co.uk
P.O. Box 241
Granger, IN 46530
Phone: 574-271-9699

TEST EQUIPMENT AND INDOOR AIR QUALITY PRODUCTS AND SUPPLIES

Air Chek, Inc.
Source of radon equipment and facts.
www.radon.com
1936 Butler Bridge Road
Mills River, NC 28759-3892
Phone: 800-247-2435

AirAdvice, Inc.
Air quality analysis systems that can be used for indoor air diagnostics for homes and businesses.
www.airadvice.com
707 SW Washington Street, Suite 800
Portland, OR 97205

Aircuity, Inc.
Manufacturer of integrated sensing and control solutions for buildings, particularly schools and commercial spaces.
www.aircuity.com
39 Chapel Street
Newton, MA 02458
Phone: 617-641-8800

Energy Conservatory
Tools for measuring pressures in and through houses and ducts.
www.energyconservatory.com
2801 21st Ave. South Suite 160
Minneapolis, MN 55407
Phone: 612-827-1117

HealthGoods, LLC
Products and information that impact the health and well-being of homes and their occupants.
www.healthgoods.com
P.O. Box 254
Derry, NH 03038
Phone: 603-434-8484

Infiltec

A great source of information and material for radon mitigation and blower door equipment.
www.infiltec.com
108 South Delphine Ave.
Waynesboro, VA 22980
Phone: 888-349-7236

Onset Computer Corporation

Full line of data loggers and data logging software for short- and long-term building performance analysis.
www.onsetcomp.com
470 MacArthur Bldvd.
Bourne, MA 02532
Phone: 800-564-4377

TruTech Tools

A wide variety of building and system diagnostic tools.
www.trutechtools.com
P.O. Box 19013
Akron, OH 443319
Phone: 888-224-3437

Appendix G
HEAT AND ENERGY RECOVERY VENTILATOR TESTING

From Bodycote Laboratories

INTRODUCTION

As residential and commercial building efficiencies are improved with insulation and improved weather stripping, building envelopes are intentionally made more air-tight, and consequently less ventilated. Since all residential and commercial buildings require a source of fresh air, the need for HRV/ERVs has become obvious. While opening a window does provide ventilation, the building's heat and humidity will then be lost in the winter and gained in the summer, both of which are undesirable for the indoor climate and for energy efficiency, since the building's heating and ventilation system must compensate. HRV/ERV technology offers an optimal solution: fresh air, better climate control, and energy efficiency.

TESTING PROCEDURES

Testing is carried out in accordance with the CAN/CSA-C439-00 *"Standard Methods of Test for Rating the Performance of Heat Recovery Ventilators."* The purpose of the tests is to determine and document the performance of a production HRV/ERV unit submitted by the manufacturer for certification.

HRVs and ERVs are tested in accordance with five test procedures contained in the CSA, C439-00 Standard, which are:

1. Airflow performance
2. Exhaust air transfer
3. Thermal effectiveness (heating mode test at 0°C [32°F])
4. Thermal effectiveness (cooling mode test at 35°C [95°F])
5. Low temperature thermal and ventilation performance (-25°C [-13°F]).

All instrumentations used for these tests are in calibration or referenced to calibrated instruments and registered in Bodycote Testing Group's Quality Assurance Program.

Prior to installing the test unit, test preparation is conducted as per CAN/CSA-C439-00, Section 10.1 to ensure that the airflow rates at the four measuring stations are within 3 percent of the average airflow value.

AIRFLOW PERFORMANCE

Airflow performance is determined by measuring the quantity of air that the HRV/ERV delivers under various external static pressures. For this test the unit fans and test loop fans (shown in Figure 1) are adjusted with dampers to provide a variety of static pressures at the HRV/ERV. The unit is normally operated at its maximum speed. The airflow rate is measured and plotted against the differences between the static pressures at the inlet and the outlet of the supply and exhaust air streams.

EXHAUST AIR TRANSFER

Exhaust air transfer is a term applied to the portion of the exhaust air stream that is transferred to the supply air stream in the recovery ventilator. The proportion of exhaust air transferred is expressed as a ratio. The term used in the CAN/CSA-C439 is "exhaust air transfer ratio" (EATR).

To measure this ratio the unit is installed in the test loop as shown in Figure 1. The supply side is configured so that uncontaminated air is brought to the cold supply inlet. The warm supply air leaving the unit is vented through an exhaust system, which discharged the air well away from the supply air inlet.

The exhaust section of the loop is closed to contain the sulphur hexafluoride (SF_6) tracer gas. The SF_6 is injected into the exhaust airside of the loop. Static pressures are adjusted to obtain the nominal settings as follows:

Static Pressures, Pa (in.w.g.)
Airflow @ Station 2 = 32.0 L/s, (68 CFM)
Station 1—Supply air entering the test unit -25 (0.10)
Station 2—Supply air leaving the test unit 25 (0.10)
Station 3—Exhaust air entering the test unit -25 (0.10)
Station 4—Exhaust air leaving the test unit 25 (0.10)

Airflow at stations 2 and 3 (supply outlet and exhaust inlet) are adjusted so that they are approximately equal in value.

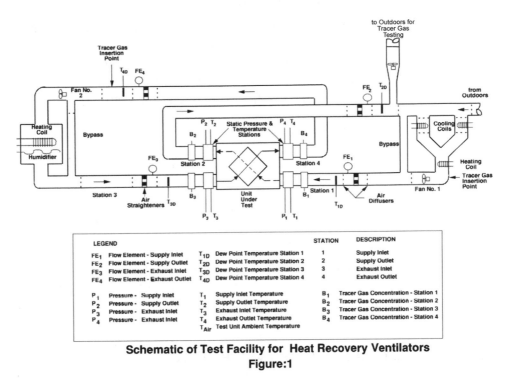

Schematic of Test Facility for Heat Recovery Ventilators
Figure:1

The injection rate of SF_6 is regulated until a steady concentration is obtained in the exhaust air stream. Tracer gas is sampled through modified flow collars installed on the supply and exhaust ports of the test unit. Alternate measurements are then taken of the supply air stream (location 2) leaving the test unit and the exhaust air stream (location 3) entering the test unit. Measurements are also taken at the exhaust air stream leaving the test unit (location 4) to ensure that a mass balance of tracer gas is achieved. The level of tracer gas is monitored using a calibrated infrared analyzer (Miran 1A). Multiple readings are taken once stable conditions are established and the averages reported for the above conditions.

SUPPLY AIR TRANSFER

Using the general testing procedure outlined above, the HRV is then tested to determine cross leakage from supply into exhaust and casing leakage from the outdoor airside. This test is described as "Supply Air Contamination" (test 2), as outlined in the CAN/CSA Standard C439-00 "Tracer Gas—Test 2" Clause 8.2.5.

The HRV is installed so that the supply section of the loop is closed, with SF_6 injected before station 1. Uncontaminated outdoor air is brought to the warm exhaust inlet. Air from the cold exhaust outlet is discharged outside of the building well away from the cold exhaust inlet.

Static pressures are adjusted to obtain the nominal settings as follows:

Static Pressures, Pa (in.w.g.)
Airflow @ Station #2 = 32 L/s, (68 cfm)
Station 1—Supply air entering the test unit -25 (0.10)
Station 2—Supply air leaving the test unit 25 (0.10)
Station 3—Exhaust air entering the test unit -25 (0.10)
Station 4—Exhaust air leaving the test unit 25 (0.10)

The concentration of SF_6 is measured at all four stations. Multiple readings are recorded once stable conditions are established and the averages are reported for the above conditions. Test 2 also compares the SF_6 concentration of the cold supply inlet to the SF_6 concentration at the warm supply outlet.

ADDITIONAL EXHAUST AIR TRANSFER TESTS

In addition to the exhaust air transfer test at the low temperature start condition, the unit is adjusted to the highest speed and the differential pressure across the core is adjusted to 50 Pa (0.20 in.w.g.) and 100 Pa (0.4 in.w.g.). The SF_6 is injected into the exhaust air side of the loop and the exhaust air transfer ratios are determined at these specific differential pressures.

THERMAL EFFECTIVENESS AND ENERGY RECOVERY EFFICIENCY

The thermal effectiveness and energy recovery efficiency of the HRV/ERV is measured in both the heating mode and cooling mode. Heating mode tests are performed with supply air temperature (station 1) maintained at a nominal average temperature of 0°C (32°F). Exhaust air conditions (station 3) were maintained at nominally 22°C (72°F), with 40 percent relative humidity.

The cooling mode tests are carried out with supply air conditions (station 1) at nominally 35°C (95°F) and 50 percent relative humidity. Exhaust air conditions (station 3) were maintained at nominally 24°C (75°F) and 50 percent relative humidity.

For both the heating and cooling mode tests, the HRV/ERV are installed in the Thermal Test Facility (Figure 1A).

The units are operated with airflows at locations 2 and 3, approximately balanced for all the effectiveness testing.

The temperatures, humidity, mass flows, and static pressures of all four airstreams, including the electrical energy consumption, are measured and recorded during these tests. These values are used in the calculation of sensible and total recovery efficiencies

LOW TEMPERATURE THERMAL PERFORMANCE

The thermal performance of the HRV under heating conditions with low supply air temperature is determined by this test. The supply air temperature (station 1) is maintained at a time average temperature of nominally -25°C (-13°F). The exhaust air (station 3) was conditioned to nominally 22°C (72°F) and 40 percent relative humidity.

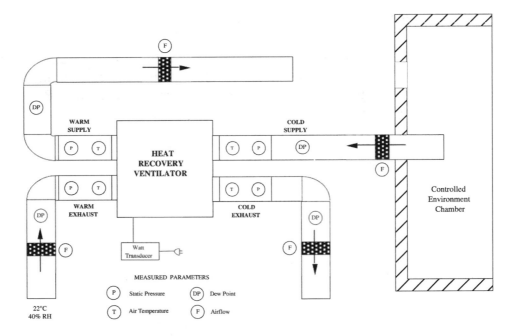

LOW TEMPERATURE TEST FACILITY SET UP

Figure 1A

For the low temperature test the unit is also set up as in Figure 1A. Testing is carried out as described in CAN/CSA-C439-00 at static pressure, airflow, and fan speed as specified by the client. Airflow rates, heat recovery, and electrical energy requirements are monitored over the 72-hour period of the test. The average values over the last 12 hours of the test are used for data analysis.

LOW TEMPERATURE VENTILATION PERFORMANCE

The low temperature performance of the HRV/ERV is determined from the low temperature test (final 12 hours of the 72-hour test). The start conditions of the low temperature test are compared to the average values of the last 12 hours of the test. From this data the low temperature ventilation reduction at maximum rated airflow is calculated as a percentage change in the maximum ventilation rate. Also, the maximum unbalanced airflow during the low temperature test is recorded as an indication of the maximum pressure unbalance potential. In addition the change in the warm supply and warm exhaust, airflows are compared to the start conditions.

SPECIFICATION SHEET

At the completion of testing a detailed report is issued outlining all the results of the above tests and a specification sheet is issued detailing the test results.

The following is an example of a specification sheet. Please note that all values in this sheet are fictitious.

HRV SPECIFICATION SHEET   Table: 2

Testing Agency:	**Bodycote Testing Group**	Model:
Date Tested:		Serial Number: -
Maunufacturer:		Options Installed:
Address:		
Telephone:		Electrical Requirements: 120 Volts 1.0 Amps

VENTILATION PERFORMANCE

Maximum Continuous Rated Airflows:			Low Temp. Ventilation Reduction Factor (CSA-C439-00)	LTVR =	0.90	
55	L/s @ 0 °C		Low Temperature Imbalance Factor	LTIF =	1.00	
30	L/s @ -25 °C		Low Temperature Ventilation Reduction During -25°C Test:		10.0	%
			Maximum Unbalanced Airflow During -25°C Test:		5.0	L/s

Airflow Range for Multispeed Units:

High Speed: L/s   Low Speed: L/s   Exhaust Air Transfer Ratio: 0.010

External Static Pressure		Net Supply Air Flow		Gross Air Flow Supply		Exhaust		Power
Pa	in. W.C.	L/s	cfm	L/s	cfm	L/s	cfm	Watts
25	0.1	72	152	69	147	74	157	85
50	0.2	63	134	63	135	68	145	84
75	0.3	56	120	56	119	62	132	84
100	0.4	49	104	49	105	55	117	83
125	0.5	44	94	44	94	49	104	83
150	0.6	37	79	37	79	44	94	81
175	0.7	32	68	29	62	37	79	80
200	0.8	26	54	23	49	30	64	78

Ext. Differential Pressure - Pascals vs Gross Airflow - L/s. Legend: ♦ Supply, ▲ Exhaust.

NOTE: FAN CURVE PERFORMED ON HIGH SPEED

ENERGY PERFORMANCE

		Supply Temperature °C	Supply Temperature °F	Net Airflow L/s	Net Airflow cfm	Supply / Exhaust Flow Ratio	Average Power (Watts)	Sensible Recovery Efficiency	Apparent Sensible Effectiveness	Net Moisture Transfer
HEAT-ING	i	0	32	30	64	1.00	46	65	81	0.00
	ii	0	32	45	96	1.02	54	63	78	0.00
	iii	0	32	55	117	1.02	84	61	73	0.00
	iv									
COOL-ING	v	-25	-13	30	64	0.98*	61	60	84	0.02
	vi	35	95	30	64	1.00	50	20**	72	0.04
	vii									

Description of Defrost: The defrost cycle, is activated when the supply air temperature goes below the manufacturers selected set point . Below set point the exhaust fan and the supply fan continues to operate at high speed. Internal dampers are activated closing the cold supply port and the other closing the cold exhaust port while simultaneously opening an internal passage that allows warm exhaust air to enter into the cold supply cavity. Warm exhaust air enters the cold supply cavity and goes through the core and out of the warm supply port. The timing sequence for this defrosts is XX minutes "off" and XX minutes "on".

Comments from Test Agency: Fan curve test was done at HRV maximum speed.

*Indicates the Supply/Exhaust Flow Ratio at 22°C prior to the start of the 72 Hour Cold Weather Test

** Indicates Total Recovery Efficiency, not Sensible Recovery Efficiency

250 Pascals = 1" of Water : 0.47 L/s = 1 cfm

Reference Report:
Sample No:

Testing was performed in general accordance with CAN/CSA-C439-00, Standard Methods of Test for Rating The Performance of Heat Recovery Ventilators, and was conducted in accordance with normal professional standards. Neither Bodycote Materials Testing Canada Inc. nor their employees shall be responsible for any loss or damage resulting directly or indirectly from any default, error or omission. Specification Sheet format revised July 2005.

INDEX